SPACE LIFE SCIENCES STRATEGIC PLAN

1992

Life Sciences Division
Office of Space Science and Applications
National Aeronautics and Space Administration
Washington, D.C.

FOREWORD

For the Life Sciences Program, the preceding year was one of significant scientific and technical accomplishments. After many years of preparation, the first dedicated life sciences Spacelab mission (SLS-1) was flown in June 1991 and yielded unprecedented information on space biology and medicine. In January 1992, the first international microgravity mission (IML-1) provided new insights into neurophysiology and gravitational biology through a set of integrated investigations conducted by a team of national and international scientists. The medical investigations conducted in support of the Extended Duration Orbiter Program provided new information on human adaptation to space flight and resulted in the prescription of countermeasures and procedures necessary to support the first 13-day Space Shuttle/Spacelab missions. Taken together, these scientific accomplishments have provided exciting new knowledge which will help us design better measures to assure human health and productivity in future, more ambitious exploration missions with humans.

The disapproval of the LifeSat series of biosatellites and the associated radiation initiative funding to support the ground-based research to address HZE biological effects has jeopardized one of the major areas of scientific investigation. Re-planning was initiated to adjust to this decision.

The NASA Administrator released "Vision 21 — The NASA Strategic Plan," dated January 1992. Vision 21 is described as NASA's "road map to the future, the NASA plan for ensuring United States leadership in space exploration and aeronautics research." The increased importance of life sciences to NASA's future was underscored.

Several other very important planning activities have taken place within the last year. In May 1991 the Synthesis Group released its report on America's Space Exploration Initiative. Also, internal and external advisory committees have made contributions to our long-term strategic planning. The latest of these, nearing completion, is a report by the Aerospace Medicine Advisory Committee. This report provides the basis for establishing research priorities and decisions necessary for life sciences programs involving human exploration missions. This activity involved the participation of more than 200 life scientists over a 2-year period.

Several significant NASA management decisions were made which included the designation of Science Centers of Excellence. For life sciences, these include the Ames Research Center for gravitational biology, CELSS, and exobiology, and the Johnson Space Center for human physiology, and operational and clinical medicine. In order to more closely integrate the Biospheric Research Program with the Earth Observing System, the transfer of most of the biospheric research tasks to NASA's Earth Science and Applications Division was initiated.

All of these activities and decisions of the last year have highlighted the role and importance of life sciences in the future of the nation's space program, both in the pursuit of fundamental scientific knowledge and in providing support to NASA's human space flight programs. The planning conducted within the last few years by the space life sciences community, coupled with the progress in our improved understanding of vital scientific and operational issues, place us in an excellent position to move toward the 21st century, to meet the challenges and opportunities that await.

Arnauld E. Nicogossian
Director, Life Sciences Division

SPACE LIFE SCIENCES
STRATEGIC PLAN — 1992
TABLE OF CONTENTS

TABLE OF FIGURES

I. INTRODUCTION

Over the last three decades the Life Sciences Program has significantly contributed to NASA's manned and unmanned exploration of space, while acquiring new knowledge in the fields of space biology and medicine. The national and international events which have led to the development and revision of NASA strategy will significantly affect the future of life sciences programs both in scope and pace. This document serves as the basis for synthesizing the options to be pursued during the next decade, based on the decisions, evolution, and guiding principles of the National Space Policy. The strategies detailed in this document are fully supportive of the Life Sciences Advisory Subcommittee's "A Rationale for the Life Sciences," and the recent Aerospace Medicine Advisory Committee report entitled "Strategic Considerations for Support of Humans in Space and Moon/Mars Exploration Missions."

Information contained within this document is intended for internal NASA planning and is subject to policy decisions and direction, and to budgets allocated to NASA's Life Sciences Program.

A.   NASA STRATEGIC PLAN

The Presidential Directive on National Space Policy, approved by President Reagan on January 5, 1988, and reaffirmed by President Bush on November 2, 1989, states that "a fundamental objective guiding United States space activities has been, and continues to be, space leadership." How best to satisfy this objective has been the subject of intense study and evaluation of possibilities and realities as NASA moves into the 21st century.

In January 1992, the NASA Administrator issued Vision 21. This plan is consistent with National Space Policy and recent major reports concerning the direction and substance of the nation's civil space program. This plan includes recommendations of the Advisory Committee on the Future of the U.S. Space Program, and the Synthesis Group on America's Space Exploration Initiative.

The essence of Vision 21 is captured in the statement of goals, missions, and enabling capabilities shown in Figure 1.

A key element of the space science mission of paramount interest to life sciences, as stated in Vision 21, is as follows:

> "Using the unique attributes of Spacelab and Space Station Freedom (SSF) to accomplish our goals in microgravity research and life sciences by facilitating fundamental advances in materials science, fluid physics, biotechnology, gravitational biology, biomedical research and long duration human space flight."

Also of great importance to the life sciences, Vision 21 contains the following direction concerning Mission from Planet Earth:

> "The first major initiative in this program, as directed by the President, is the development and deployment of SSF starting with the first element launch in 1995, man-tended capability in 1997, and permanently manned capability in fiscal year 2000. SSF enables continued progress in the human exploration of space through the prerequisite studies into human adaptation and testing of life support systems over an extended period of time. The SSF program also will give us the base of knowledge needed to build, operate and maintain large systems in space, experience that can be gained nowhere else. Freedom is indeed a critical first step of the Mission from Planet Earth and a visible symbol of America's commitment to leadership and cooperation in the peaceful exploration of space. With the establishment of a permanent human presence in space, the United States and its international partners will have, for the first time, a permanent outpost in space for performing fundamental research that will pave the way for eventual human exploration of the solar system.

1

During the coming decade, NASA will extend the duration of manned Space Shuttle flights in order to prepare for future long-duration space flights and the advent of SSF...."

Figure 1
Vision 21 — The NASA Strategic Plan

NASA'S **Vision 21** envisions an aeronautics and space program that inspires and betters the lives of all Americans, young and old, through our achievements as the world leader in space exploration and aeronautics.

GOALS

- Advance scientific knowledge of the planet Earth, the sun, the solar system, the universe, and fundamental physical and biological processes;

- Expand human activity beyond Earth orbit into the solar system;

- Strengthen the competitive posture of the United States in the fields of space and aeronautics; and

- Attract young people to the wonders of mathematics, science and technology and ensure a more technically literate society equipped for the world of tomorrow.

MISSIONS

- Space Science
- Mission to Planet Earth
- Mission from Planet Earth
- Aeronautics Research

ENABLING CAPABILITIES

- Human Resources
- Physical Resources
- Space Technology Research
- Space Station Freedom
- Space Transportation and Communications Systems

Overall, these recommendations and program direction are consistent with the traditional goals and objectives that have been pursued by NASA, the Office of Space Science and Applications (OSSA), and the space life sciences program. Over the past 5 years the relevance of the Life Sciences Program priorities and their contributions to the long-term NASA goals were examined by internal (NASA Advisory Council) and external (National Academy of Sciences) advisory committees. The findings of these reviews and resulting life sciences program formulation have placed the Life Sciences Program in a position to meet the challenges and opportunities presented.

B.   LIFE SCIENCES DIVISION OVERVIEW

The Life Sciences Division is one of eight divisions of the Office of Space Science and Applications (OSSA) of the National Aeronautics and Space Administration. OSSA is responsible for planning, directing, executing and evaluating that part of the overall NASA program that has the goal of using the unique characteristics of the space environment to conduct a scientific study of the universe, to understand how the Earth works as an integrated system, to solve practical problems on Earth, and to provide the scientific and technological research foundation for expanding human presence beyond Earth orbit into the solar system. The Division is responsible for planning, directing, implementing, and evaluating that part of the program which deals with the understanding of how living systems respond to the space environment; the search for the origin, evolution, and distribution of life in the universe; the development of the scientific and technological foundations for expanding human presence beyond Earth orbit and into the solar system; and the provision for operational medical support to all space missions involving humans.

The space life sciences program fosters research in the study of life and its processes under the influence of different environmental conditions, exhibited by both planetary surfaces and hypogravity. In order to fulfill this mission the characteristics of life-sustaining environments should be understood first, and then, by careful examination of individual parameters such as radiation, confinement, and barometric pressure changes, the reaction(s) of living organisms to hypogravity should be determined. Ultimately, by studying such processes, we will be able to document underlying critical physiological mechanisms and develop predictive models that can provide new insights about the responses of living systems to major environmental challenges. A benefit from such knowledge is the design of strategies for protecting living systems, such as countermeasures enabling humans to cope with the effects of hypogravity and safely return to Earth. Another benefit will be the development of life support systems and specific technological requirements for sustaining life in unusual and hostile environments. In the case of human space flight, the knowledge gained is transferred to the physicians and engineers for incorporation into operational systems. In the case of robotic exploration of the solar system, the information is a direct input into the requirements for the implementation of planetary protection activities.

The Life Sciences Program maintains a close working relationship with the Office of Space Flight on operational issues dealing with crew health, with the Office of Aeronautics and Space Technology in the establishment of requirements related to space technology developments, with the newly formed Office of Space Systems Development in conducting the research and development to support the operation and utilization of SSF, and with the Office of Exploration in planning for the Space Exploration Initiative. Program goals are pursued through an integrated set of activities in ground- and space-based research laboratories. The space-based laboratories currently consist of the U.S. Space Shuttle and Spacelabs, and the Russian Space Station Mir and unmanned biosatellites (Cosmos series). In the near future, the program will transition from Space Shuttle/Spacelab to the SSF; reusable biological satellites (international or U.S.-provided); and planetary probes and spacecraft. The program is accomplished with the participation of NASA field centers, other Government agencies and organizations, universities, and United States industry. Significant reliance is placed upon international participation and contributions by other spacefaring nations through established cooperative agreements.

The Life Sciences Division supports clinical and basic research. This research is conducted in ground-based, airborne, and space flight settings. The administration of the programs and projects is achieved through four closely integrated organizational elements represented in Figure 2. A more complete description of specific focused priorities for each programmatic element is presented in Appendix I (based upon the Division's recently completed disciplinary plans).

The program in biospheric research developed by the Life Sciences Division will be integrated with the Earth Science and Applications Division starting in FY 1993. The Malaria Project will continue to be directly managed within the life sciences program until completion.

In addition to the implementing structure noted above, there are three focused activities that cut across several disciplinary areas. These activities are extremely important in fulfilling our objectives.

The *Life Sciences Educational and Training Program* is an important element in developing a future cadre of expertise. The key components of the educational program are targeted toward incorporation into the high school and graduate curricula.

NASA Specialized Centers of Research and Training (NSCORT), a university/government/industry-based program, initiated in FY 1990 and focused on specific areas of interest to space life sciences. This provides for a stable multi-year funded research activity which contributes to the training of a new generation of space life scientists.

3

```
┌─────────────────────────────────────────────────────────────────────────┐
│                              Figure 2                                      │
│                 Life Sciences Division Programs/Projects                   │
│                                                                            │
│  AEROSPACE MEDICINE                      LIFE SUPPORT                       │
│                                                                            │
│   • Operational medicine                  •  Applied biomedical and life   │
│      — Clinical medicine                      support research,            │
│      — Medical standards                      including:                   │
│      — Longitudinal studies                   — Space physiology and       │
│   • Identification of requirements for           countermeasures           │
│      biomedical and life support              — Radiation health           │
│      research                                 — Environmental health       │
│                                               — Space human factors        │
│                                               — Controlled ecological life │
│                                                  support systems           │
│                                                                            │
│  RESEARCH PROGRAMS                       PROGRAMS AND FLIGHT MISSIONS       │
│                                                                            │
│   • Basic science research, including:     •  Implementation of flight     │
│      — Space biology                           projects:                   │
│      — Biospheric research                     — Aircraft                  │
│      — Exobiology                              — Space Shuttle/Spacelabs   │
│      — Planetary protection                    — Free flyers               │
│                                                — Space Station Freedom      │
│                                                — International missions     │
│                                            •  Advanced technology          │
│                                               development and              │
│                                               mission planning             │
│                                            •  Instruments for planetary    │
│                                               spacecraft                   │
│                                                                            │
└─────────────────────────────────────────────────────────────────────────┘
```

Life Sciences Research Groups (LSRGs) are being established around unique national facilities at NASA field centers. Composed of teams of scientists and engineers, they will function in a program/project mode with multi-year Research and Development activities and stable support. LSRGs utilize expertise from within NASA and from the broader life sciences community at large.

C. SPACE LIFE SCIENCES GOALS AND OBJECTIVES

The manned space program has demonstrated that Earth-bound life can be sustained beyond the confines of our planet. Unmanned planetary probes have raised many unanswered questions about the origins of life, its distribution, and the factors that have influenced its development. Through the space program the universe has become an accessible domain, opening unprecedented possibilities for probing life processes.

Following the Skylab missions primarily dedicated to life sciences in the early 1970s, the Spacelab Life Sciences missions of the 1990s open a new window of opportunity for the biomedical sciences, which will fully mature in the SSF era. Basic research is necessary to define mechanisms which are involved in the adaptive physiological changes observed in space flight and also to expand our understanding of life processes. While there is a significant overlap between the basic and applied research, it is the basic research that holds the promise for acquiring knowledge which can benefit the general population back on Earth.

The major life sciences goals are:

> - Ensuring the health, safety, and productivity of humans in space.
> - Acquiring fundamental scientific knowledge in space life sciences.

These goals in turn are supported by the following objectives:

1. To provide for the health and productivity of humans in space,
2. To develop an understanding of the role of gravity on living systems,
3. To expand our understanding of life in the universe, and
4. To promote the application of life sciences research to improve the quality of life on Earth.

The relationship of the goals and objectives to the life sciences programs is discussed below.

1. To provide for the health and productivity of humans in space

Historically, exploration of new and hostile environments was made possible through the establishment of an adequate knowledge base of the physiological responses to the prevailing living conditions and development of functional protective procedures and systems. Since humans have left the confines of Earth, it is imperative to learn how they adapt to and perform in space, and to understand the biomedical time course and health implications of space travel so that efficient life support countermeasures and medical care systems are developed, regardless of the duration of space missions.

Within the Life Sciences Division, the medical program is aimed at developing the foundations for health maintenance and care of astronauts. Some of the most important operational activities include the development of procedures to assure crew health in flight and during landing and egress of extended Shuttle Orbiter missions; development of health maintenance facilities (HMF) tailored to specific needs of different mission lengths and scenarios; definition of medical requirements for an Assured Crew Return Vehicle (ACRV); refinement of medical selection and retention standards; and establishment of medical training programs for flight crews. The primary goals of this program are to identify and anticipate potential health problems, develop preventive and therapeutic procedures, and establish research priorities to address identified biomedical challenges.

Four biomedical challenges potentially limit the duration of human space flight: physiological deconditioning; health effects of exposure to space radiation; psychosociological effects of isolated, confined and hostile environments involving complex operations through interface with space systems; and the need to meet critical life support requirements on lengthy space journeys. The research programs designed to address and mitigate these challenges are addressed below.

Space Physiology and Countermeasures concentrates on physiological decrements resulting from space flight, which become a greater concern as the duration of the space mission is extended. Ground and space research have identified unresolved scientific issues relevant to the following areas: cardiopulmonary deconditioning; neurophysiology and performance degradations, particularly space motion sickness; eliciting impacts of exposure to long-duration space flight on the immune system and the potential for illness; physiological and emotional stress and its influence on pharmacologic consequences; and changes in the bone, endocrine, and muscle systems.

Radiation poses health risks for long-duration missions in low Earth orbit and beyond, and in particular during long stays on the Moon or the 1 to 3 years required for a round trip to Mars. While considerable information is available about the physical nature of space radiation, substantive questions remain concerning the carcinogenic and other biological effects of exposure to galactic cosmic radiation and solar particle events. Additional information is needed concerning the shielding required to protect flight crews, early warning systems, and instruments to reliably measure radiation doses.

The success of long-duration missions will largely depend on the *psychosociological aspects and interactions* among the space crew and between the space and ground personnel. Little information is

5

available on interactions and productivity among small, isolated groups living in weightlessness or on planetary surfaces with a fraction of Earth's gravity level for lengthy periods of time. The most pressing issues for extended human missions, which will offer only limited possibilities for emergency rescue and return to Earth, involve interpersonal interactions, human/machine interface, crew selection, command and control structure, and crew motivation.

Environmental Factors and Life Support requirements directly relate to both the physiological and psychological well-being of space crews. The primary concerns in this area include the establishment of requirements for a regenerative food, air, and water system; development of an environmental monitoring system capable of detecting sources and types of contamination; validation of procedures for decontamination; design of the most effective systems to support EVA operations; and analysis of optimal habitability conditions for extended missions. The development of a regenerative life support system is especially challenging. The Controlled Ecological Life Support System (CELSS) Program focuses on combining biological and physiochemical processes to provide food, air, and water by recycling and removing waste materials inside the spacecraft. The behavior and productivity of plants in space, however, is not well understood, and needs to be vigorously pursued.

2. To develop an understanding of the role of gravity on living systems

Space Biology uses real and simulated weightlessness as an experimental tool to understand biological organization and function, and the role that gravity plays in the evolution and development of living systems. It addresses fundamental questions concerning how living organisms perceive gravity, how gravity is involved in determining developmental and physiological status, and how gravity has affected evolutionary history. While these questions are motivated primarily by scientific interest, such basic knowledge is critical to determining if life can function effectively for extended periods in weightlessness or reduced gravity, as on the Moon or Mars, or if an artificially induced gravitational force is required. Space-based research, which requires variable-force centrifuge facilities, provides unparalleled opportunities to expose organisms to fractional gravity levels ranging from 0 to 1 g over several generations with large sample sizes; this approach allows us to investigate the effects of gravity on living organisms in a controlled fashion.

3. To expand our understanding of life in the universe

Exobiology focuses on questions long pondered by humankind, such as: Are we alone in the universe? What led to the origin of life on Earth? Data suggest that the early environments of Mars and Earth were similar and that samples from Mars could fill gaps in Earth's geological record, in particular for the time before 3.5 billion years ago. Any valid indication of life on Mars, extant or extinct, would support the hypothesis that life can originate wherever the physical and chemical environment is favorable. For these reasons, robotic probes followed by human missions to Mars will yield important scientific answers critical to exobiology.

The *Biospheric Research Program* focuses on biogeochemical cycles and on biogenic gases as components of those cycles; studies population dynamics as the underlying mechanism of biological interactions; and constructs and validates predictive models of biospheric behavior and applies this knowledge to the study of human health. The program uses ground-based techniques to study individual sites, and extends those observations to global scale by use of aircraft and space-based remote sensing in a multidisciplinary fashion. In the near term, the program is involved in the study of globally important tropical and temperate forests and wetlands ecosystems. It is imperative that biospheric research activities be integrated with the total Earth Observing System. Therefore, the Life Sciences Division is working with the Earth Science and Applications Division in this regard, including the transfer of tasks.

4. To promote the application of life sciences research to improve the quality of life on Earth

Over the past three decades research conducted and many technologies originally developed for the space program have found important life sciences applications, which have contributed to the betterment of life for the ill, the disabled, the elderly, and the general population. Now, in the fourth decade of the U.S. space program, new benefits are anticipated from ongoing space life sciences research on the

Space Shuttle/Spacelab, and from the work to be done in the relatively near future to lay the scientific and technological foundation for future space missions.

The research conducted by the space life sciences program to date has resulted in an expansive number of "spinoffs," including significant accomplishments in the areas of:

- Understanding the role gravity plays in health and disease processes, such as
  — Orthostatic intolerance
  — Osteoporosis
  — Neurovestibular disorders
  — Muscle atrophy

- Physiological monitoring, including
  — Implantable and programmable pacemakers
  — Defibrillators
  — Medication dispensing systems
  — Intensive care monitoring procedures
  — Telemedicine

- Charged-particle therapy, especially of carcinogenesis

- Global monitoring of vector-borne diseases

- Agriculture

It is our belief that as we continue research and technology development for more ambitious missions with humans, many other benefits will become available to mankind.

D. SPACE LIFE SCIENCES PRINCIPLES

The space life sciences program is dependent upon operational experience and continuity of scientific inquiry. The scientific base is essential in planning future missions on a schedule compatible with NASA long-range options. Significant emphasis and reliance is placed upon the interactions of NASA and the scientific community in devising strategies and maintaining excellence through the peer review process outlined in the Division's Science Management Plan.

Five underlying principles are used for the implementation of a balanced and viable life sciences program:

1. Maintain and expand the unique space research facilities located at NASA's field centers, universities, and within the private sector;

2. Maintain timely and sustained access to space to conduct critical experiments;

3. Coordinate research programs with national and international organizations;

4. Develop and maintain a unique national database for space life sciences;

5. Develop and sustain an expanding training program in space life sciences.

The strategies for program implementation described in Section II have evolved within the context of the National Space Policy, NASA long-range planning, the OSSA Strategic Plan, and recommendations of life sciences advisory groups.

E. SPACE LIFE SCIENCES VISION

The second half of the 20th century has seen unprecedented breakthroughs in the fields of biology and medicine. In a relatively short period of time in the history of mankind, scientists have begun to probe life processes at the most basic level. Almost overnight an explosion in knowledge of health and disease, genetic engineering, evolution of life on Earth, and environmental biology has occurred. Despite these biomedical advances, the field of space biology and medicine still remains in its infancy. Though significant progress has been made in ground-based research, validation of these findings by actual space flight data remains problematic because of the requirements for flight opportunities.

As we approach the 21st century, we anticipate more frequent access to space for life sciences, and thus we will gather the much needed information in space biology and medicine. The legacy of current knowledge, in the Apollo, Skylab and Viking Programs has firmly established the biological and medical foundations of current knowledge in the space life sciences, and raised many unanswered questions. In 1991, SLS-1 started the revitalization of the space life sciences program. This mission produced important scientific discoveries. This highly successful mission enabled the collection and analysis of space-based research data in a wide variety of life sciences disciplines, and opened up new and important areas of investigation. This was followed in early 1992 by the first International Microgravity Laboratory (IML-1), an integrated discipline-focused international mission. The focus of investigations on IML-1 was upon plants, neurovestibular changes, human performance, radiation, and cellular differentiation.

By the turn of the century, we will have begun to understand the effect of long-duration space flight on humans, our most precious resource, allowing us to carry out an evolutionary program to study the responses of biological systems to reduced gravity and space radiation. We will conduct the research and technology development necessary to meet the needs of manned space flight programs in a timely fashion, and a longitudinal medical monitoring program to enable a full characterization of the effects of space flight on humans. There will be a natural progression in the use of Shuttle middeck, Spacelab, and eventually SSF facilities to enable biological research under space conditions. These facilities will enable human interaction with experimental systems, long-term exposure of organisms to the space environment, and the ability to react "real-time" to intermediate results in the conduct of biomedical experiments. There will be a capability to access a wide variety of organisms in space, from cells to humans. The establishment of an unmanned space research component to complement the capabilities of the currently envisioned manned space platforms will facilitate extended and repeated access to space research during periods when the Shuttle is not available to life science researchers. Such access will be provided by small satellites launched by expendable launch vehicles. We will make significant advances in understanding and quantifying the health hazards from both solar flares and galactic cosmic radiation. Studies of radiation shielding and countermeasures to physiological deconditioning will be underway to enable humans to safely venture beyond the confines of low Earth orbit and into the solar system.

During the first decade of the 21st century we will have determined and validated measures to provide medical care and preventive health maintenance, and will accurately predict the health risks for long-term sojourns in space and on the surfaces of the Moon and Mars. The provision of closed loop life support systems based on integrated biological, physical, and chemical processes, including utilization of in-situ planetary resources, will assure the capability to undertake more ambitious exploration missions.

Gravity's role in a wide variety of fundamental biological processes in plants and animals will be understood through our ability to probe mechanisms for biological sensing and transduction of gravitational forces. The systematic exploration of a wide range of gravity levels, available only through the use of suitable on-orbit centrifuge facilities, will help us understand the need for artificial gravity during long-term space travel.

Our knowledge of the relationship of life to natural processes occurring in the universe will have been expanded, and a direct search for signs of life elsewhere will have been conducted. A sophisticated microwave observing project will have completed a comprehensive search for radio signals originating from extraterrestrial technologies within a defined search space, thereby extending our knowledge of life in the universe.

By the end of this century, we will have developed and implemented the capability to predict the location, timing, and potential severity of malaria outbreaks, thereby aiding overburdened national and international health agencies around the world. Extrapolation of this capability to other diseases and ecological conditions will be underway.

Technologies developed initially for space, such as telemetry, automated fluid handling, electronics, image enhancement, and miniaturization, will have enriched conventional medical capabilities. Devices for preventive screening and monitoring, computer-aided diagnosis, diagnostic imaging tools, implantable systems, telemedicine, and research tools opening new windows into human health will continue to be the subject of productive research. In addition, we will have demonstrated a global telemedical network to aid in mitigating the effects of natural disasters.

The space life sciences program will continue to attract outstanding scientists. The recent addition of the NASA Specialized Centers of Research and Training (NSCORT) and Life Sciences Research Groups (LSRG) programs will continue to expand both the interest and contributions of the scientific community.

The life sciences program will play an important role in NASA's educational outreach effort described in Vision 21, the NASA Strategic Plan. This effort will capture students' interest in biology, medicine, and engineering at an early age, and channel students into biomedical career paths.

We expect that by the turn of this century the establishment of a national and international space life sciences entity will be a reality, benefiting not only space exploration but also continuing to stimulate scientific discovery and provide benefits to the health and well-being of all mankind.

II.   THE SPACE LIFE SCIENCES STRATEGY

A.   PROGRAM PRIORITIES

The life sciences program strategy is guided by two major goals:

Goal I — Ensuring the health, safety, and productivity of humans in space

Goal II — Acquiring fundamental scientific knowledge in space life sciences

The two goals are interactive and require continuous and careful balance. The first is mandatory in the support of overall national human space flight goals, while the second is both supportive of the first and an inherent part of NASA's space science and technology strategy. The sequence, pace, and schedules for implementation of the space life sciences program strategy are based upon the overall goals of the National Space Policy. Since 1989 the following priority order has been maintained to implement the space life sciences strategy:

1.   Maintaining the existing flight and ground research and development programs, while expanding their scope as appropriate to support future needs by the following actions:

   —   enhancing support to the NASA centers' unique facilities and talents;
   —   establishing NASA Specialized Centers of Research and Training at universities;
   —   enhancing access to space through development of new opportunities for small missions.

2.   Developing and implementing an appropriate in-space infrastructure evolving through Spacelabs, biosatellites, and SSF, and conducting critical research for advanced missions with humans;

3.   Coordinating space exploration activities with other federal agencies and international partners;

4.   Applying knowledge and technology gained from space exploration to the solution of problems on Earth.

The program strategy to provide the necessary medical and life support knowledge and scientific rationale is described in the following sections.   This strategy is consistent with the Report of the Advisory Committee on the Future of the U.S. Space Program.

B.   THE CORE PROGRAM STRATEGY

The core space life sciences program comprises ongoing programs, enhancements to the research base, small missions, and SSF utilization and support (See Figures 3 and 4). This program, while contributing to the preparation for human missions to the Moon or Mars, would be needed even in the absence of an overarching initiative in order to further the goals of NASA and the space life sciences community.

The strategy for the acquisition of fundamental scientific knowledge in space life sciences must meet two challenges. In the first challenge, the program must be responsive to the priorities of the two primary communities it serves — the scientific community and mission implementors. The latter group seeks applied knowledge in specific disciplines to enable the development and implementation of missions with optimal life support, countermeasure and medical care systems. The scientific community seeks to utilize space flight opportunities to obtain basic and applied scientific knowledge. Because of the nature of the life sciences, a very high degree of commonality exists in the research priorities that have been established by both communities.

The second challenge is to match science with the most appropriate platform. A significant portion of the life sciences research can be conducted in ground-based laboratories. However, this research, in order to be fully understood, requires that the ultimate validation of hypotheses and ground-based experimental models be conducted under space flight conditions. When flight experiments require humans as either the most appropriate test subjects or as the experimentors, then the manned mission is the necessary

10

```
┌─────────────────────────────────────────────────────────────────────┐
│                             Figure 3                                  │
│                        Current Programs                               │
│                                                                       │
│  •   Research and Analysis              •  Operational Medicine       │
│      — Space Physiology and Countermeasures   — Extended Duration Orbiter Medical │
│      — Space Human Factors                    Program (13-Day)        │
│      — Environmental Health                                           │
│      — Radiation Health                 •  Flight Programs            │
│      — Controlled Ecological Life Support System   — Space Shuttle/Spacelabs │
│        (CELSS)                              -  Space Life Sciences (SLS) mission │
│      — Space Biology                           series                 │
│      — Exobiology                           -  International Microgravity Laboratory │
│          -  Search for Extraterrestrial Intelligence   (IML) mission series │
│             (SETI)                          -  Cooperative international missions │
│      — Biospheric Research              — COSMOS/MIR                  │
│      — NASA Specialized Centers of Research and   — Life Sciences Laboratory Equipment │
│        Training (NSCORT)                — Space Station Freedom Utilization │
│      — Life Sciences Research Groups (LSRG)    -  Crew Health Care System (CHeCS) │
│                                             -  Space Biology Initiative │
│                                             -  Centrifuge Facility    │
└─────────────────────────────────────────────────────────────────────┘
```

```
┌─────────────────────────────────────────────────────────────────────┐
│                             Figure 4                                  │
│              Life Sciences Planned Programs Initiation                │
│                                                                       │
│           Enhancements to Research Base           Flight Missions     │
│                                                                       │
│  1993   •  Life Sciences Data Archiving System                        │
│                                                                       │
│  1994   •  Research Community Revitalization    •  Explorer-class series │
│         •  Extended Duration Orbiter Medical Program  •  Biomedical Monitoring and │
│            (16-Day)                                Countermeasures (BMAC) Program │
│                                                                       │
│  1995                                           •  SSF Research Testbeds │
│  through                                        •  EDO Spacelab series │
│  2002                                                                 │
└─────────────────────────────────────────────────────────────────────┘
```

platform. When scientific objectives can be achieved without direct human intervention, then an unmanned spacecraft is the preferred platform. When the most appropriate setting is a ground-based laboratory, the strategy focuses upon the use of university and/or industry facilities, unless NASA-unique capabilities are required.

To meet these challenges, **five themes** have shaped the life sciences strategy for NASA. The first theme addresses the priorities of the mission implementor community and supports the vision of progressively extending human ability to cope with and live productively in the space environment:

THEME 1. Conduct focused research and development on the ground and In space to develop the knowledge In medical and life support arenas to meet the requirements of NASA's current and future manned space flight programs.

This theme results in research and technology necessary to meet the needs of manned space flight programs in a timely fashion, and a longitudinal medical monitoring program to enable a full characterization of the effects of space flight on humans. As a result of its operational component, the Life Sciences Division leads all human support activity conducted within NASA, either by setting requirements or by direct investment.

Humans have successfully lived and worked in the weightless environment of space — occasionally for lengthy periods of time. Through observation, subjective accounts and data collection, we have learned that adaptive physiological changes occur that may reduce performance in flight and impair the ability to readapt to a gravitational environment. It should be noted that these data have usually been collected while countermeasures were in use. Little information exists from only a few individuals who did not use countermeasures or used them only in a partial way. One of the conclusions which can be made from these observations, however, is that the current generation of countermeasures is only partially effective. These countermeasures have reduced some symptoms during short flights. However, the longer missions planned for the future will require significant improvement in these countermeasures in terms of ease of use and efficacy. The countermeasures currently used address the symptoms, not necessarily the causes.

The *Extended Duration Orbiter Medical Program (EDOMP)*, initiated by the *Operational Medicine Program* in 1989, is designed to develop specific medical countermeasures for the extension of Space Shuttle missions to 13 days. The validation for 13-day missions will be accomplished on United States Microgravity Laboratory-1. The knowledge acquired as a result of EDOMP activities will feed into an evolving database to support more advanced missions on SSF. The extension of Spacelab missions will enable experiment replication on the same flight, thus increasing the scientific and operational return from each mission. An extension of the EDOMP program to enable 16-day Shuttle/Spacelab missions will be the subject of a new initiative in FY 1994.

To be implemented in time to support SSF missions, the *Biomedical Monitoring and Countermeasures (BMAC) Program* will develop countermeasures for cardiovascular, bone, muscle and neurological disorders associated with progressively increasing space flight durations. The BMAC Program will also monitor the physiological status of crews during routine on-orbit operations so that the efficacy of the countermeasures can be determined, the individual crew person's physiological status can be followed, and the individual crew person's optimum productivity achieved.

The goals of the BMAC Program will be to:
* provide systems and procedures that optimize crew performance on orbit, and
* minimize postflight time required for crew members to return to flight status, and
* evolve knowledge base for future mission planners.

The BMAC Program will be implemented in successive phases to match the evolving requirements during the development of SSF (SSF) capabilities, including continuing Long-Duration Orbiter (LDO) support.

The BMAC Program will work closely with other life sciences activities on SSF. The SSF *Crew Health Care System (CHeCS)* will provide both clinical support (diagnosis and treatment) for crew members in the event of injury or illness, and operational monitoring of the SSF internal environment to assess compliance with environmental standards that affect crew health. The basic biomedical research requiring human subjects that is to be done on SSF by NASA, other federal agencies (such as NIH) and by our international partners will be coordinated with the BMAC Program. Such coordination is necessary so that there is no conflict or interference between the countermeasure protocols required by BMAC and the desire on the part of the biomedical community to do research requiring human subjects on SSF.

Additional strategic themes address the means by which the space life sciences conducts its research in space and satisfies the requirements of the scientific community it serves. The next two themes address the efforts to match mission features and capabilities with scientific requirements for access to space.

THEME 2.   **Develop laboratory capabilities in space that will enable long-duration, state-of-the-art, biological and medical experimentation to be accomplished.**

This theme results in a natural progression in the use of Shuttle middeck, Spacelab, and eventually SSF facilities to enable biological research under space conditions. These facilities will enable human interaction with experimental systems, long-term exposure of organisms to the space environment, and the ability to react in "real-time" to intermediate results in the conduct of biomedical experiments. This theme also results in a requirement for access to and maintenance of a wide variety of organisms in space, from cell-culture systems to humans.

Scientific research in space has been costly to date, partly because it is often necessary to build new research equipment for each flight experiment, and partly because experiments have to meet stringent limitations on weight, size, electrical power consumption, and environmental conditions. Therefore, for life sciences experiments on Spacelab Life Sciences (SLS), International Microgravity Laboratory (IML), and other Spacelab missions, investigators coordinate their research to minimize the need for hardware duplication. The implementation of this strategy to date has resulted in the NASA space life sciences programs developing flight experiments for seven unmanned COSMOS biosatellite missions with the U.S.S.R., Space Shuttle middeck locker experiments, and a variety of Spacelab missions, both dedicated to the U.S. life sciences community and in concert with other national and international partners.

The continuing Spacelab mission series now enable the performance of experiments for up to 13 days on-orbit. A range of analytical capability is provided on these missions, with the primary objective being to perform observations and collect samples and specimens during flight for subsequent Earth-based analyses. The initial Spacelab Life Sciences mission (SLS-1) was successfully conducted in June 1991, followed by IML-1 early in 1992. Many investigators use NASA *Life Sciences Laboratory Equipment (LSLE)*, an inventory of multipurpose, reusable medical and biological research facilities developed or modified for use in space. While most of the current hardware is quite old, LSLE replacements and upgrades will provide much needed state-of-the-art capabilities and be ready for transition to space station operations. A *life sciences data archiving system* is being implemented to facilitate the orderly preservation and dissemination of integrated data sets and specimen samples. The archive system, as currently envisioned, is planned to be accessible through a common network providing controlled access to all investigators. The archive will be available in time to house the myriad data sets derived from the SLS-1 and IML-1 missions and is planned to evolve to meet the data management challenges of the SSF era.

Small missions currently in the detailed planning stage include a continuation of the extended-duration *Spacelab series* with research conducted in nearly all space life sciences disciplines. The primary and secondary emphasis and schedule for each of the planned major space life sciences missions are shown in Figure 5.

Space Station Freedom (SSF) has great significance for and represents significant challenges to the NASA space life sciences programs. SSF gives us the opportunity to conduct life sciences investigations over extended periods of time to: 1) separate the effects of natural aging processes from those of hypogravity; and 2) to understand the effects of hypogravity on multiple generations of living organisms. For the first time it will be possible to replicate experimental data from a variety of simultaneously exposed species with appropriate controls and real-time analytical capabilities over extended periods of time. At the same time, a system of

Figure 5
Major Space Life Sciences Flight Opportunities

DISCIPLINE	DESCRIPTION	SLS-2 (1993)	SLS-3 (1995)	NEUROLAB* (1997)
CARDIOVASCULAR	Cardiopulmonary, fluids, electrolytes	1	2	
MUSCULOSKELETAL	Calcium metabolism, endocrine-hormonal metabolism, muscle function	2	1	
NEUROSCIENCE	Vestibular physiology, neurobiology	2		1
REGULATORY PHYSIOLOGY	Endocrinology, metabolism, immunology, pharmacology, renal physiology, rhythms	2	2	2
SPACE HUMAN FACTORS	Human-machine interactions, performance	2	2	2
MEDICAL CARE TECHNOLOGY	Surgical workstation, health maintenance facility, diagnostics		2	
GRAVITATIONAL BIOLOGY	Plant biology, developmental biology			2
EXOBIOLOGY	Gas-grain simulation		2	

1: PRIMARY DISCIPLINE EMPHASIS
2: SECONDARY DISCIPLINE EMPHASIS

* PLANNING STAGE

monitoring and ameliorating the physiological adaptations that occur in humans subjected to extended space flight must be evolved to provide the continuing operational support to the SSF crew. Freedom will provide the means to acquire basic knowledge on mechanisms of gravity perception while paving the way for extended-duration exploration missions with humans.

A strategy for space life sciences research activities that are of high programmatic and/or operational relevance, are technologically feasible, and can optimally support and utilize the capabilities to be provided by all phases of an evolving international SSF has been developed and under implementation since 1989. This strategy ensures full operational support to SSF crew activities while enabling significant scientific research during the evolutionary process of facility development and deployment of on-orbit capability over a 5-year period (1996 - 2001). Each phase builds upon the preceding infrastructure and enhances the capability to perform research in support of a wider range of science requirements.

The research phase of the *Life Sciences Space Station Freedom (LSSSF)* Program will commence with the utilization or outfitting flights following the deployment of the U.S. laboratory module and achievement of Man Tended Capability (MTC). This phase will be marked by the evolutionary implementation of a research infrastructure to establish an initial national and international life sciences research laboratory on SSF. The NASA LSSSF program is currently developing a suite of discipline-focused "common-core" facilities, the *Space Biology Initiative,* capable of supporting the variety and range of research efforts to be undertaken by the Life Sciences community on SSF. These generic facilities have been defined by science working groups derived from both internal and external communities and in close consultation with our International Partners. It is the intent of the Life Sciences Division to implement a research infrastructure on SSF through which scientific investigations, openly solicited from the scientific community, will be realized. Experiment-unique or experiment-specific equipment required by each peer-reviewed investigator will be developed on an "as-needed" basis and sequenced into the research timeline as available resources (logistics, rack volume, crew time) become available to the discipline.

The *Centrifuge Facility* for SSF is expected to be the single most important research tool for space life sciences and will provide a number of critical capabilities. The facility will provide suitable and appropriate environmental habitats furnishing life support for nonhuman experimental species. It will provide the means to produce artificial gravity and accurately controlled acceleration levels in a microgravity environment. It will enable the unambiguous and thorough execution of studies addressing the many important critical questions in a number of life science disciplines by providing one-gravity control populations. It will provide a testbed for determining the interplay of artificial gravity and biological specimens within a controlled spacecraft environment. Such experiments will allow the precise analysis of the effects of microgravity and radiation as well as varying levels and exposure times of linear acceleration on biological systems. In addition, the Centrifuge Facility will provide a critical tool for a significant number of pioneering experiments requiring small animals and plants in both acute (short term, minutes to hours) and chronic (long term, days to months) experiments.

Specifically, the centrifuge will make possible three space research capabilities that were heretofore not available. These unique capabilities can be considered the scientific purpose for the Centrifuge Facility:

• A 1-g control environment in order to separate the effects of microgravity from other environmental and physiological factors. Such an "environment" dictates that life support capabilities and environmental qualities be common and controlled between experimental and control subject populations.

• A means of conducting fundamental biological studies in which gravity, the parameter of interest, is variable and under experimental control. These include studies of the minimal level (thresholds of intensity and duration) of gravity required to maintain normal structural and functional mechanisms from the cellular to the organismic level and among the different physiological systems (e.g., cardiovascular, musculoskeletal and regulatory physiological systems). A means to expose specimens to two or more gravity levels simultaneously is required for such experiments. Studies of exposure to gravity levels up to 2 g will address artificial gravity as a possible countermeasure for long-duration manned flight.

• A source of 1-g conditioned specimens that are also adapted to the spacecraft environment prior to experimental use, and which are available at any time during a space mission.

To date, only very small centrifuges (a fraction of a meter in diameter), accommodating only a limited range of small specimens, have been flown on Space Shuttle/Spacelab missions and on Soviet biosatellites and space stations. The Centrifuge Facility will provide a major technological advance to enable investigations on a variety of test subjects over durations representing all or a significant portion of an organism's life-span.

The Life Sciences Division has invested significant resources in planning for operations of a fully outfitted *International Life Sciences Research Facility* on SSF. Such a facility is planned to support continuous scientific investigations for more than 20 years and meet the requirements for:

- Research devoted to in-depth study in each medical and biological discipline over dedicated periods of time (up to 6 months);
- Establishing a capability to address medical issues which will enable future exploration missions with humans.

The NASA Life Sciences Division has established an *International Life Sciences Strategic Planning Working Group (ILSSPWG)* encompassing each of our International SSF partners. One of the primary functions of the ILSSPWG will be to develop and maintain a "common-path" approach to the provision of "generic" facilities for and the implementation of discipline-focused science across SSF increments. The ILSSPWG will develop a consensus implementation schedule for increment-specific life science research on SSF and promulgate focused Research Announcements on a regular basis, soliciting research by individual discipline area (e.g., Plant Biology, Cardiopulmonary Physiology, Exobiology, etc.), for specific SSF increment intervals.

Each regular, periodic science recruitment will be open to the entire International life science community, will be fully peer-reviewed and endeavor to offer a "rapid" response capability to the implementation of scientific experiments within the resources and infrastructure available to the life sciences on SSF within two years of each recruitment. It is anticipated that the initial discipline-focused Research Announcements will be released approximately three years prior to the establishment of Man-Tended Capability on SSF and will be widely promulgated as the first Research Announcement in a continuing, regularly scheduled series of such discipline-focused recruitments.

THEME 3. **Ensure frequent, assured, access to the space environment for life sciences investigators across the entire range of the biological and medical sciences that can be advanced by space research.**

This theme requires both the use of frequent middeck experiment opportunities on the Shuttle and the establishment of an unmanned space research component to complement the capabilities of the currently envisioned manned space platforms. The space life sciences need extended and repeated access to unique orbits that cannot be reached by the Shuttle program or SSF, and access to space research during periods when the Shuttle is not available to life science researchers. Such access will require the use of small satellites launched by expendable launch vehicles such as the seven successful COSMOS biosatellite missions with the U.S.S.R. Figure 6 shows the planning schedule for launching all small class missions involving life sciences.

To replace the *COSMOS* program and to continue the access to space flight, a series of *Explorer-class mission* operations, that enable studies using living biological specimens including plants, rodents, cell and tissue cultures, and other small organisms, is in the planning stages. In addition to providing easy, frequent and inexpensive access to space , these mission operations will have a unique set of technical capabilities, including access to orbits that the Space Shuttle cannot reach (e.g., polar orbit), extended flights (e.g., 30 to 60 days), and exposure to varying gravity levels (e.g., 0 to 1.5 g). Dedicated space biology research and focused radiation studies will be the prime subjects of the Explorer-class missions to begin in 1998. The strategy under development will include ground-based spacecraft simulations, robotic technology for executing radiobiology experiments remotely ("telebiometry"), radiobiology of large-scale cell cultures required for low dose and dose rate experiments with a high degree of statistical significance, development of hardware required for long-term life support of large number of small mammals similarly required for statistical precision, radiobiology of small mammals particularly suited to exploit space flight data including interactions with weightlessness, and the use of

16

Figure 6
Life Sciences Mission Opportunities
Calendar Year Quarters

	1	2	3	4
1992	IML-1		SL-J	COSMOS
1993	SL-D2	SLS-2		
1994	IML-2			
1995		SLS-3	SL-D3/E1*	
1996				
1997	SL-E2* SSF*	Neurolab* SSF*	SSF*	SSF*
1998	SSF*	SSF*	SSF*	SSF*
1999	SSF*	SSF* Biosatellite*	SSF*	SSF*

*Planning Stage

available flight opportunities on the Shuttle, on SSF and on international collaborations (e.g., Eureca 1997). The current program calls for definition studies in FY 1992-1993 and initiation in FY 1994, with a decision on the actual platform(s) to be used for verification (lunar radiobiological laboratory, BioExplorer or other alternatives) to be made in time for initiation FY 1998.

THEME 4.   Use NASA capabilities to search for and understand evidence of life and life-related molecules in the universe.

This theme results in the use of solar system exploration spacecraft to conduct *in-situ* research in exobiology, in SSF facilities to conduct experiments on prebiotic chemistry, in the use of NASA's Great Observatories and other spacecraft to look for clues to the origin of the biogenic elements and compounds, and in the use of NASA facilities to conduct a ground-based search for evidence of extraterrestrial technologies.

The *Search for Extraterrestrial Intelligence (SETI) Microwave Observing Project*, a 10-year program to probe our galaxy for radio signals of possible extraterrestrial intelligent origin, was initiated to address one of the major questions of the Exobiology Program. Using existing radiotelescopes in NASA's world-wide Deep Space Network and additional telescopes made available by the National Science Foundation and foreign organizations, a targeted search of nearby solar-type stars and an all-sky search will be conducted which will be more comprehensive than the sum of all previous searches. The technology needed to mount the search was developed within the base program, and already appears to have applications outside the space field. Deployment of the initial SETI systems is planned for 1992, with full operations in 1995.

THEME 5.   Apply techniques and understanding gained through experience in space to the alleviation of human suffering on Earth.

This final theme brings the products of space research home in a fundamental way. Space provides us with an opportunity to better understand biology, medical systems and practice, and the functioning of the living Earth as a system. NASA's life sciences program seeks to transfer this understanding and experience to help those who will never, themselves, journey into space.

Underlying all the work in support of the five themes are strong, continuing *Research and Analysis programs* including <u>applied life support and biomedical research</u> (Space Physiology and Countermeasures, Space Human Factors, Environmental Health, Radiation Health, Controlled Ecological Life Support Systems); and <u>basic scientific research</u> (Space Biology, Biospheric Research, and Exobiology).

In order to revitalize the academic community's contributions to the space life sciences, a number of *NASA Specialized Centers of Research and Training (NSCORTs)* are being established. The first five centers are focused in gravitational and developmental biology, environmental health, bioregenerative life support, exobiology, and radiation health. The NSCORT Program is an integral part of the Division's research and analysis program to advance basic knowledge in the space life sciences and generate effective strategies for coping with, and eventually solving, space life sciences problems in focused research areas. The program is expected to further the Nation's scholarship, skills, and performance in the space life sciences and related technological areas and enhance the pool of research scientists and engineers trained to meet the considerable challenges inherent in the Nation's commitment to prepare for future human space exploration missions. The NSCORT Program is designed to mobilize talent and other resources in academia and industry, with meaningful participation from scientists in NASA and other Federal agencies, and to derive the benefits from concentrated administration and financing in a research and training environment.

A new approach to consolidate and focus NASA field center life sciences activities has been formalized, allowing the designation of *Life Sciences Research Groups (LSRGs)* charged with the responsibility of integrating the research activity of a number of tasks into a coherent, peer-reviewed, research program. In 1991, the first LSRG was formed at the Johnson Space Center in the area of bone and muscle research. The LSRG designation implies that the team is a national resource in a well-defined area of the space life sciences, is involved in a wide spectrum of basic and applied research tasks, and possesses an appropriate set of laboratory facilities and the scientific expertise to accomplish their agreed-upon tasks. LSRGs may have unique research facilities associated with their laboratory. If this is the case, it is expected that such facilities will be operated as national facilities, and will be available for use by the external research community, including other Federal agencies.

C.   STRATEGY FOR MISSION FROM PLANET EARTH

This section describes augmentations to the core program that will be required to enable the Mission From Planet Earth (MFPE). While no single mission set with clearly enunciated schedules has been selected, initial planning has determined mission destinations, established priority sequences for mission implementation, defined general mission objectives, and identified key programs and supporting elements. The sequence of SSF, return to the Moon and evolution to self-sufficiency, followed by human exploration of Mars, defines the nature of the critical areas for research efforts to establish life sciences/life support requirements.

Human missions beyond low Earth orbit pose an enormous challenge to the space life sciences. While the augmentation to the elements of the core programs described above are essential to any such undertaking, additional major issues also must be addressed. The proposed missions to the Moon and Mars, ranging in length of up to more than 3 years, will expose crewmembers to a number of known and unknown hazards. The key issues of uncertainty involve the effects of exposure to radiation and extended stays in microgravity and reduced gravity environments. More efficient life support systems to provide closure of air and water loops and food production, including *in-situ* utilization of resources, are needed. Another critical component is adequate medical care and health maintenance support. Human factors considerations such as crew selection, interpersonal relations, human-machine interface, and habitability will also be pivotal in the ability of the crews to function effectively and safety.

During the latter part of FY 1991, a major planning activity was initiated. The Aerospace Medicine Advisory Committee (AMAC) was requested to develop an implementation strategy in support of human exploration missions. The life sciences research program developed during the last few years has addressed a diverse set of important science and mission operation considerations. The AMAC undertaking addressed the need to utilize and enhance this existing program in a manner which ensures that results are obtained in support of a focused and integrated program of exploration. Such a program is required to provide timely answers for the high risk critical questions in order to proceed safely with the exploration mission(s). This required adopting an exploration oriented perspective and evaluating the critical questions from the 12 life sciences disciplines. What are the most critical issues? Are there constraints that will prevent establishing a lunar outpost or the undertaking of a human mission to Mars? When is the critical window by which the information has to be known? What are the acceptable levels of risk? What is our current level of scientific and technological readiness? The resulting plan, summarized below, describes the proposed life Sciences implementation strategy for exploration missions. Specific areas of research, schedules, milestones, and deliverables are addressed and prioritized, including the use of Spacelabs and SSF to develop the required knowledge base. Implementation strategy will be paced to the national agenda.

1.   Environmental Health and Life Support Systems

One of the major unknowns for any human mission outside the Earth's magnetosphere and for exploration and habitation of the Moon and Mars surfaces is the *radiation health risk*. Specifically, the level of biological damage produced by galactic cosmic radiation (GCR) and solar particle events (SPEs) must be quantitatively assessed. GCR is the major source of continuous radiation outside the magnetosphere. Even though GCR has a relatively constant intensity and is fairly well-characterized, it has the highest uncertainty level with respect to risk assessment. This is due mainly to the large uncertainties of the biological effects from primary and secondary charged particles. The all-pervasive GCR burden represents a long-term excess cancer mortality risk. Currently, there is no appropriate model for early or late biological effects from this radiation.

The SPEs are lower in energy and easier to shield against than the GCR. However, they are highly variable in energy and intensity from event to event, and for anomalously large events represent a lethal, acute space radiation hazard. There is at present no adequate model that can predict either the occurrence, energy, or relative intensity of SPEs.

MFPE requires a definitive risk assessment of space radiation hazards and enabling science validation through ground- and space-based research. This research program is necessarily complex, since it requires a multidisciplinary integrated approach in solar physics, nuclear physics, theory (modeling), radiobiology, and probability risk assessment. Impacts due to radiation issues on spacecraft design, habitats, and mission planning must be assessed with a high level of confidence to meet MFPE milestones. Robotic precursor Mars missions, with sample return capability, will provide significant information on the SPEs and physical environmental conditions that will be encountered by humans.

In the radiobiology area, structured activities have been initiated to explore research to assess radiation risks and determine how best to manage those risks. The implementation of these efforts entails an interagency effort among NASA, DOE, and DOD research programs in order to utilize existing facilities and scientific manpower in a timely manner. The enabling science and technologies require ground- and space-based facilities and programs which have 3 to 5-year lead times to be in place and fully operating.

While an expanded ground-based program could develop a great majority of the scientific basis for predictions of space radiation hazards and shielding requirements, these predictions are based on simulating the space radiation environment with a relatively small number of particles, at a restricted number of energies. Before such predictions can be used for the management of space radiation risks, they need to be validated in the space environment, where all particles, at all energies, are incident from all directions. Thus, the space experiments are critical for a full verification of the predictive models.

Conducting operations in space requires that provisions be made for protecting people from its inhospitable environment. Living organisms possess a remarkable degree of adaptability, but humans can only survive in an environment characterized by rather narrow thermal and atmospheric limits. Furthermore, adequate food and water are required to sustain life. The consumables needed to sustain life can either be brought from Earth or, with the right technology, created in place from wastes or *in-situ* resources. This tradeoff between resupply, regeneration, and manufacturing is embodied in the concept of "loop-closure." U.S. space programs to date have employed open-loop system design, with no reuse of waste products (e.g., carbon dioxide, waste water streams, biological wastes), which are stored for return to Earth or are vented overboard from the spacecraft. However, the type of closed-loop systems critical to the success of the exploration program have never been developed. Creation of closed-loop life support systems based on regeneration of waste products represents a radical departure from existing experience.

This function can be partially accomplished by physico-chemical life support systems alone, which can primarily recycle water and air and remove contaminants from the environment. Incorporation of the biological element into life support systems, however, can add another dimension for loop closure, insofar as food production is concerned. Plants are also extremely efficient and reliable processors of air and water. This is the basis for the *Controlled Ecological Life Support System (CELSS)*.

A CELSS integrates biological and physico-chemical processes to construct a system that will produce food, potable water, and a breathable atmosphere from metabolic and other wastes, in a stable and reliable manner. Such systems include plants for air revitalization, potable water processing, food production, and a waste processing system to convert human and plant wastes into acceptable water, CO_2, and nutrient solutions for the plants. The key research areas involved in the development of a CELSS include plant growth techniques, food production, waste processing and contaminant removal, and system integration and control. While full-scale life support systems based on bioregeneration are the ultimate objective of this developmental activity, limited applications of bioregenerative techniques can yield significant benefits as well. For example, plant growth units could provide fresh salad vegetables, partial air revitalization, partial water purification, and a significant psychological benefit to crewmembers engaged in tending and eating these plants. Applications for this activity would include lunar surface habitats and orbiting lunar laboratories, as well as a Mars transfer vehicle or Mars habitat.

The expanse and success of the exploration program will be dependent upon the ability to perform routine *extravehicular activity (EVA)* in a variety of gravitational environments. Existing technology (Apollo and Shuttle suits) was designed to work in specific environments which are not readily applicable to exploration requirements. The design drivers for use in microgravity and on planetary surfaces are such that different suits might be necessary. The configuration of suits (and the degree of commonality) for use in microgravity, and in lunar or Martian environments, will be determined after trade studies are conducted. As a pacing element for exploration, it will be essential that an advanced EVA development effort be established. Since lunar missions will occur a decade or so before an excursion to Mars, the initial focus of such a program should be an operational lunar suit and developmental prototypes of a Mars suit for testing on the Moon.

Establishment of environmental standards and development of technologies for real-time environmental monitoring and environmental decontamination will be required. The development of compatible *intravehicular and extravehicular activities life support systems* with long-term efficiency and reliability represents a significant technological challenge for the space life sciences. Because of the criticality of life support and the technical immaturity of the relevant subsystems, an accelerated developmental program is necessary in this area. This program must focus efforts on development of both subsystem and integrated system technologies, as well as a core research program evaluating applicable physico-chemical processes. This effort will benefit from work done on SSF life support systems.

2. Countermeasure Systems

Gravity has shaped (or influenced) the evolution of most living things on Earth. Adaptation to microgravity represents the body's response to the lack of certain stresses, using mechanisms that were shaped in a terrestrial setting. The Skylab Program permitted the first detailed studies of this adaptive process, and the Soviet space program has collected data for stays of approximately 1 year on orbit. The U.S. and Soviet programs have shown that space flight exerts a significant and complex influence upon human physiology, leading to a varied range of adaptive (and maladaptive) responses, which if not curtailed by the use of countermeasures, can lead to medical problems. These biomedical manifestations may vary in nature according to the duration of exposure and type of countermeasures used. This process of adaptive responses, termed "space flight deconditioning," poses risks for compromising mission objectives, if not adequately controlled and mitigated.

Changes in human physiology must be understood to modify the course of space flight deconditioning and enable the delivery of medical care in space (acute medical care decisions are often based on changes in underlying physiological indices). Satisfactory resolution of the medical and technical unknowns in this area will require an active program of ground-based studies, as well as inflight investigations. To date, our experience with long-duration missions, comparable in duration to the length of time to reach Mars, is quite limited. Figure 7 depicts the cumulative number of crewmembers who have

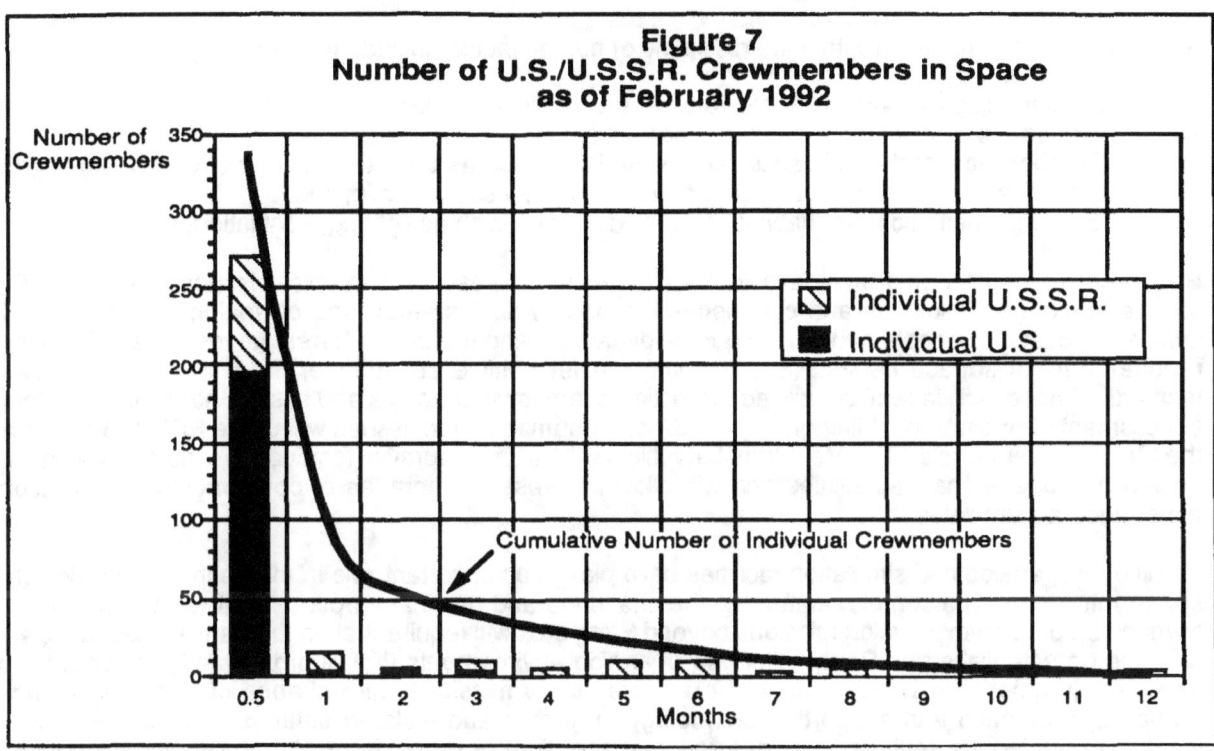

Figure 7
Number of U.S./U.S.S.R. Crewmembers in Space as of February 1992

flown missions of different duration. The majority of U.S. and U.S.S.R. astronauts have flown missions of less than 15 days in length, thus attesting to the inadequacy of currently available biomedical data on long-duration missions which could lead to the development of modern countermeasure systems.

Another concern involves the effects of partial gravity on a crewmember already deconditioned by exposure to microgravity. The significant commitment of time and resources which will precede the initial Mars landing makes it imperative that the crew have adequate physiological reserve to perform assigned duties. While readaptation to Mars gravity (0.38 g) is not anticipated to be as pronounced as return to the Earth's 1-g environment today, the understanding of fractional-g physiology is essentially nonexistent, preventing an assessment of the relative medical risks.

If the deconditioning resulting from extended stays on planetary surfaces is not adequately counteracted, even minor negative trends, operating over extended periods of time (such as 600 days on Mars), would have the potential of resulting in physiological decompensation on return to 1-g. It is critical that the physiology of deconditioning and readaptation to various g environments be sufficiently understood to assure that the crew do not suffer serious medical consequences upon return.

If exercise and other forms of countermeasures should prove inadequate or operationally infeasible, artificial gravity may be required. Consequently, in addition to the continuation of an expanded Biomedical Monitoring and Countermeasures (BMAC) type program, a parallel research effort to define requirements for artificial gravity systems should be conducted.

3. Human Factors

The field of human factors encompasses a broad range of disciplines which primarily address the research areas relevant to human interface with a variety of aerospace systems. Anticipated activities in the MFPE program will expose crewmembers to a unique combination of stresses and hazards for sustained periods of time. Effective integration of human factors requirements into mission design results in two primary accomplishments: a) the human will be physically and mentally able to do the tasks outlined; and b) all the systems, equipment, spacecraft, rovers, vehicles, tools, etc., will be designed such that the tasks can be accomplished. The final outcome of a well-designed system is a synergy between the human operator and the system.

Specific concerns contained within the discipline of human factors include the following:

- Habitat design, including usable volume, space allocation, etc.;
- Human-machine interface considerations, such as workstation design;
- Psychological and psychosocial considerations, such as crew selection, motivation, small-group dynamics, provisions for recreation, and optimal crew size and mix; and
- Environmental and other habitability considerations such as lighting, ventilation, etc.

Research elements would include a focused effort to address psychological issues (crew selection, effects of extended isolation and confinement, group dynamics, etc.), use of analog environments for evaluating design concepts and studying crew dynamics, and the use of terrestrial testbeds. The design requirements of surface habitats and a Mars transfer vehicle are unique and will require dedicated testbeds. These facilities will be utilized for testing operational protocols and system hardware, and for an assessment of habitability. Microgravity validation of human factors issues will utilize SSF and potentially the Mars transfer vehicle. The Mars transfer vehicle will be a microgravity testbed to refine and validate the design and operational capabilities and will allow progressive upgrades of components as operational experience accumulates.

Traditionally, analog and simulation facilities have played an important role in the definition and design of space missions. The complex nature of the challenge and the many options that will be available as humans embark on exploration missions beyond Earth orbit will require that, in the early stages, simulation facilities be established on Earth. The use of analog environments that approximate in important ways those environments to be encountered on future space missions will enhance our understanding of human performance and supporting technology in remote and isolated settings. Simulation facilities located in analogous settings to long-duration space flight or outposts on the Moon and Mars include undersea habitats and Antarctic research sites.

4. Medical Care Systems

Within the U.S. space program, inflight medical illness has resulted in only minor mission impacts, presumably due largely to the good health, relatively young age of the astronaut corps, and the limited length of missions. In contrast, the Soviet space program, whose missions are often lengthy, has been affected on occasion by inflight medical contingencies, resulting in either a mission abort or significant replanning. While predicting the probability of medical illness or inflight injury is difficult, it is reasonable to assume that medical contingencies can occur at any time in the course of an ambitious and sustained exploration program.

The approach to inflight medical care is one of risk management. Since it is not cost effective to duplicate the entire capability of a modern hospital onboard a spacecraft (or surface habitat), the general principles underlying terrestrial medicine should be adhered to, within the context of existing program constraints. Crew health can conceivably be affected by a number of factors, including: a) preexisting medical risk; b) physiological deconditioning, which if not corrected may result in or aggravate a medical event; c) occupational injuries due to inherently hazardous operations; and d) risks secondary to spacecraft environmental factors, especially those due to failure or degradation of life support systems.

The primary considerations affecting the design of an inflight medical care system include the degree of remoteness from Earth (and subsequent limitations on medical transport), nature of operations, degree of hazard involved, crew size, mission length, rescue capability, and degree of acceptable risk. Eventual development of an effective medical care system for exploration will be dependent upon advances in several key areas. The principal areas of concern include the following:

- Delivery of medical care in space will be dependent upon an understanding of space-flight physiological deconditioning and upon the development of procedures and protocols for clinical care in reduced and microgravity environments.

- In addition to developing an adequate understanding of space-adapted physiology, knowledge of how the body responds to pathological conditions (i.e., shock or trauma) in microgravity will be required to practice space-based medicine. This data will build upon an understanding of the "normal" space-adapted physiology, and will be accomplished by means of a focused research program utilizing SSF.

- Adaptation of "off the shelf" medical technology will be the principal hardware thrust; however, there are certain key capabilities that the program will need. These include computer-aided medical diagnosis, extended-life pharmaceuticals, blood substitutes (or freeze-dried blood), and compact imaging and diagnostic systems. Advances in these areas would be significant in terms of their applications within the medical field.

Significant progress in the development of inflight medical care systems can be made by use of the Space Shuttle as a platform for the development and validation of medical hardware. Modest, sustained effort will allow early implementation of terrestrial medical testbeds, joint development of certain critical medical technologies with the SSF Program, development of medical care procedures/protocols utilizing "zero-g" aircraft, and development of a university-based telemedicine developmental laboratory to refine space telemedicine applications. Studies to clarify overall medical care requirements for various missions and hyperbaric treatment will be included.

5. Outposts on the Moon and Mars

Although the specific pace and implementation plans are not yet defined, NASA's preliminary approach to the development of outposts on the Moon and Mars consists of four phases. The first, robotic exploration, obtains data to assist in the design and development of subsequent human exploration missions and systems, demonstrates technology and long communications time operational concepts, and dramatically advances scientific knowledge of the Moon and Mars. The second phase, outpost emplacement, emphasizes accommodating basic human habitation needs, establishing surface equipment and science instruments, and laying the foundation for future, more complex instrument networks and surface operations by testing prototypes of later systems. The third phase, consolidation, further expands these capabilities, and the fourth phase, operation, entails a steady-state mode with the maximum possible degree of self-sufficiency.

Space life sciences research to support human exploration will progress incrementally as the program proceeds. In the early stages, SSF will serve as a controlled testbed for studying extended-duration human habitation in space and for developing and validating systems and elements, such as habitation and laboratory modules and life support systems, to be used later on the Moon and Mars. Preparation for human missions to Mars will require a series of robotic missions after the Mars Observer to support and verify landing site selection, identify hazards to human explorers, and prepare for science experiments conducted by the crew.

SSF will be utilized to perform enabling research and to gain operational experience of importance to the MFPE. Of particular interest will be the BMAC Program and the scientific research made possible through the use of the Centrifuge Facility and the facilities provided by the Space Biology Initiative.

As with all elements of Mission from Planet Earth, *in-situ* science will become progressively more sophisticated as the program proceeds. On both the Moon and Mars science capabilities should begin with local human exploration complemented by unmanned rover traverses and be followed by the emplacement of initial science instruments. Later, more advanced research facilities can be built, including pressurized life sciences testbeds to be used for basic and applied life sciences research. Upon the initiation of the emplacement phase of the lunar outpost, an additional focus for life sciences research will be on systems developed on the Moon itself. Early systems will be used to establish prototypes for long-term habitation, and later habitats will provide additional space for increased biomedical and life sciences research. The facilities will be used to simulate the eventual long-term stays anticipated for a Mars mission. Later, space life sciences research in preparation for Mars missions can be conducted at the lunar outpost. Advanced medical and life support technology development of systems to protect and support human space travelers must also be conducted. Areas of concern include radiation protection, reduced gravity countermeasures (including artificial gravity), medical care, life support (including EVA), and resolution of behavioral and human factors issues.

There are critical gaps in our understanding of exobiology that can only be filled by space exploration, including missions to the Moon and Mars. Exploration missions to the Moon will permit the collection of solar, interstellar, and interplanetary dust particles that will reveal the cosmic history of the biogenic elements and compounds. Simulations of dust-grain chemistry, studied in the lunar environment, will further our understanding of the synthesis of complex organic molecules in space. Exposure of terrestrial organisms will allow extensive studies of the survivability and adaptation of life in the space environment. Perhaps the most important benefits will be the use of the lunar surface as a platform for observations of organics in planetary atmospheres and interstellar environments. Analyses will be performed on samples from the polar regions and the deep subsurface for possible prebiotic molecules delivered by cometary impact, while samples from a variety of lunar craters will significantly improve our understanding of the impact of events that have affected the evolution of complex life on Earth. Using the lunar far side to search for radio signals from intelligent species beyond the solar system will extend similar terrestrial observations into domains inaccessible from Earth.

Exploration missions to Mars will provide a unique opportunity to understand the role of life in the evolution of the terrestrial planets. The role of life in determining planetary evolution is one question whose answer could have immediate consequences in understanding global change on Earth. On Mars, exobiologists will seek to understand whether life ever arose, and if not, why not? Identifying differences between chemical evolution on Mars and Earth will enhance our understanding of life's origins. If there was life on Mars, then studying the fate of that life and its possible survival into the present will be the major science driver for the exploration of the planet.

There is considerable evidence that Mars had liquid water on its surface at one time. It is possible that life originated on Mars during a wetter early period. Understanding the history of water on Mars is a key requirement for finding sites in which biological activity may have occurred, and in which a record of that activity is recorded. Early robotic missions are critical for addressing the question of extant life on Mars and for laying the foundation for studies of the existence of life during an earlier period. Mars Observer data will reveal information about water, mineralogy, and the structure of the planet. Instruments with an analytical capability to measure evolved gases, electrical properties of the soil, and chemical, elemental, and isotopic composition need to be developed. Future missions will also require instrumentation to search for and select samples for terrestrial analysis for extinct and extant life, including imaging devices, mass spectrometers, and increasingly capable sample acquisition devices.

Earth-based research will permit effective understanding of the data returned from the exploration missions. Terrestrial analogs of the Mars environment (e.g., the Antarctic, simulation chambers, etc.) will be used to develop microanalytical instruments and laboratories for exobiological analysis of returned samples, to better understand the chemical properties of the soil and its biological potential, to learn how to detect fossils in sediments that once harbored life, and to apply the lessons learned at Mars to our questions about how life originated on Earth. Young scientists who will go on to address these questions must be trained to enable future work on the Martian surface. Manned operations on Mars will allow analysis of elements, organics, oxidants, and gases, and extensive microscopy to significantly enhance the search for signs of life. Iterative, real-time interaction between the scientist and Mars may be essential to a full understanding of Mars exobiology.

D.   STRATEGY FOR SUPPORT TO MISSION TO PLANET EARTH

The Biospheric Research Program combines space flight and airborne observations with ground-based research to determine how living and nonliving elements of the Earth are linked by physical, chemical, and biological processes. These processes are integrated over the entire planet by the land, atmosphere, oceans, and sediments to form a system called the biosphere. The core Biospheric Research Program is dedicated to understanding how biological and planetary processes interact, and how, in conjunction with the environmental effects of human activity, these processes are affecting the long-term habitability of the Earth.

The Biospheric Research Program has developed a new research initiative, *"Terra."* The goal of this initiative is to arrive at an understanding of population and community dynamics using remotely sensed data. Terra was developed as the Life Sciences Division's contribution to NASA's overall program in Earth System Science. The planned program will exploit remote sensing technology, including Earth Probes and the Earth Observing System (Eos), to increase our understanding of ecological interactions. The major objectives of Terra are to:

- Continue and extend present studies in biogeochemical cycling,
- Support a new emphasis on research at the population and community level using remote sensing,
- Enhance support for applications of population-level studies to disease-vector populations,
- Promote development of remote sensing expertise in the ecological community.

Through Terra and associated activities, the Biospheric Research Program will implement the goals noted in the Ecological Society of America's sustainable biosphere initiative, and the reports "Earth System Science: A Closer View" (NASA Advisory Council) and "Global Change in the Geosphere-Biosphere" (International Geosphere-Biosphere Programme).

The Biospheric Research Program has also undertaken the planning and coordination of the Freshwater Initiative, a cooperative interagency program being organized under the framework of the Federal Coordinating Council for Science, Engineering and Technology-Committee on Earth and Environmental Sciences. The Freshwater Initiative is designed to acquire a predictive understanding of freshwater ecosystems and resources that can be used to improve detection, assessment, and prediction of environmental effects, and to develop effective management approaches and mitigation alternatives for potential global change scenarios. The leadership of the Freshwater Initiative has been assumed by the National Science Foundation.

Biospheric Research Program activities will be closely integrated with the Earth Science and Applications Division activites starting in FY 1993.

E.   DECISION RULES FOR INTEGRATING OVERARCHING INITIATIVES

The process for establishing priorities and sequencing the missions that contribute to overarching initiatives is determined by the thematic relationship the space life sciences goals and objectives have to each initiative and the budget level available in support of the initiative. Though some programs will be refocused to support such initiatives, preservation of the core science program vitality will require an enhancement of the resources historically available to the program. The implementation of missions dedicated to the Mission From Planet Earth (MFPE) will require resources beyond the core program level. The strategy to expand elements of the core program that make significant contributions to MFPE, such as the Space Biology Initiative, the Biomedical Monitoring and Countermeasures Program, and the Space Radiation Health Program, will be dependent upon NASA emphasis and availability of resources, matching funding pace to implementation schedules. Without such enhancements, these elements will remain in the science- or mission operations-driven, integrated queue.

In undertaking significant new mission initiatives in support of MFPE, the Life Sciences Division will assess their relevance to and impact upon the balance of the overall science program and enabling themes, and the capability of the institutional infrastructure and the scientific community to accomplish them.

The decision rules are:

1.   Match the implementation schedule of mission initiatives to the pace at which NASA and the Nation proceed.

2.   Establish an implementation schedule compatible with the technological maturity of the mission and with the infrastructure capability to deliver.

F.   ACCOMPLISHMENTS IN 1991

The year 1991 was one of significant accomplishments for the Space Life Sciences Program. The major highlight was the flight of Spacelab Life Sciences 1 (SLS-1), the first Space Shuttle mission dedicated to life sciences research. The SLS-1 mission was accompanied by progress across the broad array of life sciences activities.

Spacelab Life Sciences 1 (SLS-1) Mission

SLS-1 was a milestone in NASA's ongoing effort to gain an integrated understanding of physiological adaptation to space flight. Future efforts will build upon the knowledge obtained on this mission.

SLS-1 was launched on June 5, for a 9-day mission. While previous flight investigations established the general physiological changes in space, research on SLS-1 was primarily devoted to understanding the underlying biological processes and mechanisms of adaptation to microgravity and readaptation to Earth's gravity. The crew participated as subjects and scientist-operators. For the first time in the history of the space program, an integrated set of investigations was conducted utilizing multiple biological subjects ranging from single cells to humans.

The mission was an outstanding success. Nearly all of the planned data collection was accomplished, with minimal problems experienced with the over 3500 hardware items flown. The primary payload consisted of 18 investigations — 10 utilized human subjects, 7 utilized rodents, and 1 utilized jellyfish. The experiments were developed by an international team consisting of 17 principal investigators and 35 co-investigators from four countries — Australia, Canada, Switzerland, and the United States. The mission's success was due in large part to the dedication of the crew. Almost an entire day of additional, unscheduled scientific activity was gained through the crew's interest and initiative. Crewmembers participated in anthropometric, cardiovascular, pulmonary, neurovestibular, musculoskeletal, blood, human factors, and metabolic studies. Among the new techniques and procedures initiated on SLS-1 were tracers for measuring fluid volumes, and state-of-the-art automated cardiopulmonary tests. A number of physiological measurements were taken for the first time in space.

SLS-1 resulted in more life sciences data than any previous mission. Due to its sheer volume, however, the data will take months of analysis and interpretation. Data collection was started earlier in flight than ever before, to capture those changes that begin with the onset of weightlessness. Postflight experiments were initiated immediately after landing, providing a unique opportunity to study the adaptive processes from launch through landing as a continuum.

Significant scientific results were obtained in areas that have interested researchers for years. Many of the findings were anticipated, confirming theories and validating ground-based tests. Investigators reported a decrease in body weight in humans and animals due to fluid and tissue losses, decreased red cell mass, orthostatic intolerance following flight, degradation of anit-gravity muscles, initial symptoms of bone demineralization, and immune suppression of lymphocytes.

Even more dramatic were the unanticipated findings, which compel us to reassess our ideas of how individual body systems adapt to space flight. These findings include evidence that adaptation begins earlier than heretofore expected and that the nervous and endocrine systems play a significant role in regulating that adaptation. SLS-1 provided new insights into the fluid shift phenomena, including a greater degree of initial plasma volume loss than expected, a more rapid change in central venous pressure than was thought possible, and an elevation of renal clearances that did not diminish during the flight as previously thought. These insights will stimulate a reassessment of ground-based models of fluid shift. Assessment of pulmonary function revealed the totally unexpected finding that regional differences in pulmonary ventilation and perfusion, previously thought to be a totally gravity-dependent phenomenon, still exist in weightlessness. Iron uptake studies have suggested new mechanisms to explain the loss of red cell mass in space. Preliminary analysis of rat vestibular tissue suggests an increase in neural synapses, a significant finding that was entirely unexpected, but suggests that adaptive changes can fundamentally affect the nervous system. An *in-vitro* study demonstrated that under certain conditions cellular (lymphocyte) activation may be considerably enhanced in space — a result that may have exciting implications for biotechnology. Advances in electron microscopy and biocomputational data analysis are being applied to the reconstruction of three-dimensional neural networks from micron-thin vestibular tissues from rats flown on SLS-1.

Data collected on SLS-1 are providing background information necessary to understand physiological adaptation to weightlessness, and will contribute to the preparation for future long-duration space flight missions. Lessons learned from SLS-1, both scientific and operational, will be incorporated into planning the SLS-2 payload and future SLS missions, including improvements in techniques and protocols. SLS-2, scheduled to be launched in mid-1993, will provide science enhancement and statistical significance for the data collected on SLS-1.

Extended Duration Orbiter Medical Program (EDOMP)

The Extended Duration Orbiter Medical Program (EDOMP) is directed toward ensuring Space Shuttle crew health and safety on missions of up to 13 days.

The Space Shuttle differs from previous spacecraft in that it is manually piloted through the final landing phase. The crew, seated upright, is subjected to gravity stress (up to 1.2 g's) for 15 to 17 minutes, and throughout this time they must be able to perform precise landing operations.

The major operational issues being investigated are maintenance of the pilot's physiological capacity for effective performance during re-entry and landing, the assurance of adequate muscle strength for unaided emergency egress, and the determination of any potential occupational health risks associated with the extended mission. The latter issue involves the monitoring of gas composition and airborne particles in the Shuttle cabin. In addition, possible changes in neuromuscular function related to eye-hand coordination during landing are being studied.

In 1991 a major addition to the program was the acquisition of two Crew Transport Vehicles for carrying the crew as rapidly as possible after landing to a medical facility for data collection and medical examinations. Rapid transport is essential, as readaptation to Earth's gravity begins immediately upon reentry.

Like the results from SLS-1, the knowledge acquired through the EDOMP will feed into the overall database to support the lengthier missions planned for SSF. The first 13-day Shuttle flight is currently planned for United States Microgravity Laboratory 1 (USML-1) in 1992.

New Spacelab Life Sciences Missions

Advances in technology have revealed more about the brain and nervous system in the last 10 years than we have learned throughout history. To recognize this progress and stimulate new advances in the neurosciences, President Bush and Congress have designated the 1990s as the "Decade of the Brain." Initial planning commenced in 1991 for one of NASA's primary contributions, "*Neurolab*," a Spacelab Life Sciences mission planned for the 1997-1999 time frame. Neurolab will be wholly dedicated to the neurosciences and behavioral sciences, allowing investigators to use the unique environment of weightlessness to study the function and development of the nervous system, and the systems that sense gravity and control posture and movement. Initial plans call for research on communication and sensory disorders, and behavioral adaptation to space flight. Other potential research areas include studies of neuronal function at the cellular level, and vestibular and proprioceptive physiology. Neurolab will utilize a new generation of research equipment. All of this hardware will have the capability of being transitioned to SSF, to enable longer term studies of the central nervous system in space.

Planning for "*Ecolab*" was also in the initial stages in 1991. Ecolab will be a Spacelab mission dedicated to the systematic examination of the role of gravity and other environmental factors on ecological interactions. The mission will investigate contributions of the effects of microgravity and closed systems on biological systems. Investigations will focus on how the absence of gravity affects such processes as microbial growth, photosynthesis, plant growth, metabolism, decomposition, mineralization, and the production and adsorption of gases.

Space Station Freedom

Significant advances were made during 1991 in the planning and development for SSF, which will provide the first opportunities in the U.S. space program since Skylab in the 1970s for life sciences investigations over extended periods in space, with on-orbit controls and analytical capabilities. During the year decisions were made to make life sciences research the primary reason for developing the space station.

Life sciences research on the station will have two principal themes—operational and basic scientific research. In many cases, these themes overlap; an enhanced understanding of the mechanisms underlying adaptive responses to weightlessness will help us refine the countermeasures that will enable long-duration exploration missions.

The 2.5 meter centrifuge facility to be installed and operated on the space station completed its competitive Phase B studies in February 1991. A major program review followed, the overall concept and technical feasibility was accepted by the review panel, and weaknesses were identified and addressed, solidifying the technical requirements and programmatic approach. Life sciences program scientists reviewed those plans in September 1991 to ensure they were compatible with science requirements from all the life sciences disciplines and to provide additional planning information. These activities lead to the release of the Phase C/D Request for Proposals (RFP) in the spring of 1992.

The Biomedical Monitoring and Countermeasures Program was the subject of preliminary planning leading to definition studies in 1992.

Planning for the Gravitational Biology Facility resulted in formalization of the program management structures and the assignment of key personnel. Major activity centered around preparations for a Level I Requirements Working Group meeting in October 1991.

Key CELSS Test Facility scientific, technical, and management documentation reached increased levels of maturity by the end of 1991. Also, the CELSS Test Facility Science and Technology Working Group was formed and the first meeting was held in October 1991. Similar progress was made in planning for the Gas-Grain Simulation Facility.

Research and Analysis (R&A)

A major accomplishment during 1991 was the completion of science plans in each of the life sciences disciplines. This was based upon the thorough review of each discipline by Discipline Working Groups, each composed of several outstanding scientists from the space life sciences community.

In the Space Physiology and Countermeasures Program, continuing research supported during 1991 the preliminary identification of further research needs and priorities and operational requirements for extended-duration flight missions. Data collected during the SLS-1 mission are proving extremely valuable in this regard, and planning for subsequent missions is underway.

In the Toxicology discipline of the Environmental Health Program significant progress was made in developing spacecraft maximum allowable concentrations (SMACs) for potentially hazardous substances. Toxicological assessments were conducted for four Shuttle missions, work continued on NASA's toxicology data base, and contingency plans for toxicological events in space were developed based on this improved information. In the Barophysiology discipline, an effort continued during 1991 to define and verify risks for decompression sickness (DCS) and to develop protocols that preclude DCS without extensive pre-breathe periods prior to EVA. Other studies to deal with DCS also are underway, with a long-term goal of understanding the mechanisms of tissue fluid homeostasis in microgravity and hypoxic environments. In the Microbiology discipline, studies are underway to minimize the possibility for flight crew infection through the analysis of spacecraft air, water, and surfaces, and the definition of acceptable standards.

Primary 1991 activities in the Exobiology Program involved preparing for the observational phase of the SETI microwave Observing Project (1992), the Mars Observer mission (1992), the U.S.S.R. Mars '94 mission (1994), the arrival of the Galileo mission at Jupiter, the launch of the CRAF and a Cassini missions to a comet and the Saturn system, and the installation of the Gas-Grain Simulator on Sapce Station Freedom. Steps were also being taken to prepare for future Space Exploration Initiative missions to Mars and later the Moon. One early step is planning and the initiation of joint research with the National Science Foundation (NSF) in Antarctica in the 1992-93 season.

In 1991, a workshop conducted at the Johnson Space Center involving over 100 principal investigators in radiation health facilitated the discussion of their work and the exploration of new directions in space radiation health. Also during 1991, the LifeSat Program progressed through Phase B studies and was in planning for Phase C/D contracts before being terminated by Congressional action. Other activities included interagency planning for cooperative research and the use of major non-NASA facilities in radiobiological research.

The Space Human Factors Program made an important outreach to the scientific community in 1991 in the planning of cooperative research projects. For example, joint efforts are being discussed with the National Oceanic and Atmospheric Administration to use NOAA's undersea habitat to study psychosocial interactions and other human factors issues.

One of the Life Sciences Division's central goals is the acquisition of fundamental knowledge concerning space biological sciences. Investigations conducted by the Space Biology Program use microgravity as a tool to advance knowledge in the biological sciences, understand the role of gravity in biological processes, and identify how different species are affected by and adapt to microgravity. An active program of flight experiments and ground research has continued with a study of circadian rhythms in space in the fungus Neurospora, conducted on STS-32 in 1990, and experiments investigating muscle metabolism in rats on STS-32 in 1990 and on STS-48 in 1991. Space biology experiments in neuroscience, the musculoskeletal system, and cell biology were key investigations on SLS-1.

The CELSS Program is developing a bioregenerative life support system utilizing a combination of biological and physical/chemical processes. The goal is to construct a biologically based system incorporating fresh crop plants, that can produce and recycle food, air, and water in a safe and reliable manner. It is anticipated that the presence of green plants in an otherwise wholly artificial environment will have aesthetic and psychological benefits as well. An ongoing activity over the last year was the investigation of the effects of lighting and temperature on crop growth in a Crop Growth Research Chamber at Kennedy Space Center (KSC) in Florida. Also at KSC, the CELSS Breadboard Project continued to investigate different plant crop yields.

An ongoing important activity is the development of an operational database for archiving results of the NASA life sciences research program—past, present, and future—for dissemination in a readily accessible form to the life sciences community nationwide, and eventually, internationally. Data from space and ground-based studies will be accommodated in its various forms—digital and analog, imagery, fluid samples, tissues sample, etc. The long-term goal is to develop a multi-mission archival system, along with a master index of all publications containing reports of space life sciences research.

Two additional NASA Specialized Centers of Research and Training (NSCORTs) were established in 1991. To date five centers have been established: Bioregenerative Life Support (Purdue University), Gravitational and Developmental Biology (Kansas State University), Environmental Health (University of Rochester), Radiation Health (Lawrence Berkeley Laboratory), and Exobiology (University of California, San Diego).

Operational Medicine Program

A successful year of Space Shuttle missions has enabled the collection of significant quantities of flight crew physiological data to aid in the planning of subsequent Shuttle missions and the assessment of physiological factors affecting future missions of longer duration on Extended Duration Orbiter and SSF.

Mission From Planet Earth

During 1991, progress was made in better defining the physiological and environmental factors potentially limiting the duration of human space flight and the relationships of those factors to the various planning options being considered for missions to the Moon and Mars as part of the Space Exploration Initiative (SEI). Communications were improved significantly among those within and outside the life sciences program who are involved in SEI planning.

G. ANTICIPATED PROGRESS IN 1992

The Life Sciences Division's plan for FY 1992 was discussed in the 1991 issue of the Strategic Plan. Budgetary decisions contained in congressional legislation for FY 1992 necessitated reductions in several parts of the life sciences program. The revised plan is discussed below.

1. Flight Programs

Several *Space Shuttle Spacelab missions* will be supported in FY 1992. Among these was International Microgravity Laboratory 1 (IML-1), of which approximately 50 percent of the payload related to space life sciences. The focus of U.S. activity on IML-1 included studies of plants, neurovestibular changes, human performance, radiation and cellular differentiation. The IML-1 mission was conducted in early 1992 in a very satisfactory manner.

Work will continue on the *Extended Duration Orbiter Medical Program (EDOMP)*. The first extended duration (13-day) Shuttle flight is currently planned for USML-1 in 1992. By that time, approximately 100 Detailed Supplemental Objective (DSO) experiments will have been conducted on missions of different durations to determine acceptable parameters for maintaining crew health and safety on USML-1 and future extended-duration missions. Development of exercise hardware will be completed in FY 1992 and inflight evaluations will be conducted leading to alternative countermeasures. An improved anti-g suit and operational procedures will undergo flight evaluation and the new Baseline Data Collection Facility will be completed and utilized.

In FY 1992, development of upgraded replacements for the *Life Sciences Laboratory Equipment (LSLE)* for Space Shuttle use will accelerate, as many items of existing hardware, some in use since the early 1980's, are phased out.

Efforts will continue on definition and development of experiments previously selected through the Announcement of Opportunity (AO) or NASA Research Announcement (NRA) process and focused on hardware that will be flown on several future Spacelab/Shuttle missions in FY 1992 and beyond, i.e., Shuttle middeck/secondary payloads, the Japanese SL-J mission, the second dedicated life sciences mission (SLS-2), the German D-2 mission, IML-2, and SLS-3. Collaboration with the Russians on the *COSMOS* biosatellite program will continue with joint research on a COSMOS flight in 1992. Additionally, opportunities will be pursued to utilize the Soviet Mir space station for some research activities and hardware verification. Detailed definition and hardware development will continue on an integrated centrifuge facility for SSF.

Activity will continue in the *Space Biology Initiative* to determine how biological research can be accommodated on SSF, as well as to define generic instrument and facility requirements. Studies will identify unique scientific and hardware transition requirements from continuing Spacelab flights to SSF operations. In addition, technology assessment, advanced technology development, hardware design and development, and experiment definition and planning will be performed.

2. Research and Analysis (R&A)

The research and analysis activity supports space life sciences program goals of advancing knowledge in all areas of space life sciences and developing medical and life support systems that enable human habitation in space. Funding restrictions in the life sciences budget are having particularly adverse effects on R&A activities. Planned enhancements in several important areas are being delayed to stay within smaller than anticipated budgets.

The **Space Physiology and Countermeasures Program** will focus on the study of chronic problems associated with extended durations in space. Information will be collected on occupational exposure in microgravity on each Shuttle flight and potential countermeasures will be evaluated, especially in the areas of vestibular dysfunction, cardiovascular deconditioning, and musculoskeletal atrophy. This information will be used in the research program to optimize crew performance and to develop countermeasures for extended manned missions.

The **Space Human Factors Program** will continue the development and testing of procedures, protocols, instrumentation, and equipment for future space exploration. Guidelines for human-machine interactions and automation requirements will be established; a dexterous EVA glove will be designed using laser mapping technology; and crew selection and support systems, intervention techniques, and coping regimes for stress management will be further developed.

The **Environmental Health Program** will focus on environmental risk assessment by developing a toxicology database and contamination modeling capacity, determining spacecraft maximum allowable concentration (SMAC) limits, and developing new technologies for environmental monitoring. The program also will focus on characterizing the spacecraft environment by assessing air quality and the effects of laboratory operations, and by determining the presence of irritant gases.

Research in **radiobiology** requires the continuing development of instrumentation that complements traditional dosimeters and is capable of characterizing energetic charged particles. Research also focuses on determining optimum shielding and other countermeasures to ameliorate the consequences of radiation exposure. Studies of the mechanisms of radiation action at the cellular level will continue to be pursued, and attempts to elucidate their relationship to effects in animals and humans will continue from a systems perspective. The program will also use available studies of human populations accidentally exposed to radiation in order to refine radiation limits already established for low Earth orbit.

A key element of space life sciences enabling research is **CELSS**. Near-term CELSS research will emphasize system definition studies for a hybrid physico-chemical-biological life support system intended for long-duration space missions. The program will also support critical studies in controlled-environment plant production, food processing and nutrition, and regenerative waste processing.

The **Space Biology Program** will continue to focus on the mechanisms by which calcium ions and calcium-mediated physiological mechanisms are affected by gravity. Biochemical and physiological processes which regulate growth and development in plants and animals will be studied on the ground, under conditions that simulate physiological changes produced in weightlessness, and compared to actual flight experiment results. The program will continue to expand the knowledge of cell division rates, chromosomal integrity patterns of differentiation, and changes in cell shape as they are affected by exposure to weightlessness. Experiments to address these issues will be conducted in the Shuttle middeck (small payloads) and in Spacelabs. Additional experiments will be flown on the Soviet biosatellite mission and possibly on the Space Station Mir.

The **Exobiology Program** will, in the nearterm, emphasize the development of new flight experiment concepts for future use on planetary exploration missions and to use SSF to investigate models of early solar system evolution. The program will also support research on the mechanisms for the synthesis of biologically significant molecules in space, their incorporation into planetary bodies, and the origin and subsequent evolution of life on Earth.

The **Biospheric Research Program** will continue to support research leading to an understanding of how biological processes interact with planetary processes and how humans affect this interaction. Specifically, ground-based research and models of biogenic gas production in wetlands, temperate forests, and tropical forests will provide the baseline necessary to expand to the regional and global level using remote sensing technology. Another area which will receive continued emphasis is the detection and monitoring of disease vectors (specifically the mosquito that carries malaria) using a combination of ground-based and remote sensing techniques.

3. Search for Extraterrestrial Intelligence (SETI)

The *SETI Microwave Observing Project* is planned to begin operation in October 1992 to coincide with the 500th anniversary of Columbus' arrival in the New World. The primary activity in the project during FY 1992 will be to continue building and testing the system in preparation for initial full systems deployment. The first of the Targeted Search systems will be completed, integrated, and tested in FY 1992. Field testing and later integration into the Arecibo telescope in Puerto Rico will take place. The purchase and assembly of the remaining systems will be started, and planning will be continued for their deployment to observing sites worldwide. Other activities will include the development of operational planning for both the Sky Survey and the Targeted Search.

4. NASA Specialized Centers of Research and Training (NSCORTs)

The initiation of the NSCORT Program and the establishment of the first centers in the latter part of 1990 were highly significant steps in the strengthening of university-based research in space life sciences. In FY 1992, planning for two additional NSCORTs will be conducted. They will complement the five existing NSCORTs established in 1990 and 1991, leading over the next few years to the implementation of an NSCORT in each major space life sciences program area.

5. Mission From Planet Earth

Beginning in FY 1992, where resources permit, the space life sciences program will conduct Pre-Phase A and Phase A studies in areas focusing on participation in human and precursor lunar-Mars missions. These areas include artificial gravity, planetary protection, advanced remote medical care, and human factors research. Trade studies of physiological effects of microgravity and engineering options for artificial gravity systems will be a significant component of this activity.

Priority of activities pursued will be based upon results of the AMAC study and subsequent NASA management decisions.

III.   IMPLICATIONS OF THE SPACE LIFE SCIENCES STRATEGY

Fiscal year 1989 was a pivotal year in the space life sciences program. The development of the science implementation strategy with community consensus has paved the way for the establishment of a cohesive life sciences program. This program is designed to optimize utilization of resources to achieve formulated goals and objectives. In 1989 we began the Extended Duration Orbiter Medical Project as an enhancement to the Operational Medicine Program. This enhancement was driven by NASA needs to extend the flights of the Space Shuttle. The definition of the Centrifuge Facility Project also began in 1989. In 1990 we provided for an increase to the U.S./U.S.S.R. cooperative efforts in biology and medicine through participation in the Mir program. We also initiated the Phase B studies for the Centrifuge/Space Biology Initiative and began to accelerate the SETI Microwave Observing Project. SETI Phase C/D and the NSCORT enhancement to the base Research and Analysis programs got fully underway in 1991. Phase B of the LifeSat platform as part of the Radiation Biology Initiative (RBI) was concluded in 1991. The base program remained at a level-of-effort and the delays in the Spacelab Life Sciences missions diverted resources to maintain schedule readiness, with no modern flight hardware added for subsequent missions. While the overall program has shown a healthy growth in 1990 and 1991, major problems remain in the implementation of enabling research needed to support SSF.

In 1992, space life sciences is faced with another austere budget, resulting from Congressional action on the FY 1992 budget request, reducing or eliminating funding in several areas. Direction was received to:

— delete $15 million from the LifeSat Program, terminating the program;
— reduce life sciences flight experiment funding by $5 million; and
— reduce $15 million from life sciences research and analysis, to be primarily derived from activities associated with Moon-Mars efforts.

The formal FY 1993 budget request has been submitted to Congress, however, action is not yet known.

The implementation of the present program was made possible by adhering to the following budgetary decision process since 1989:

1.   Maintain the core research and development program for:

   a.   near-term approved flight facilities and experiments (those in design and development, Phases C/D);
   b.   research and engineering infrastructure at the NASA field centers through the establishment of program/projects; and
   c.   viable university and industrial base. This base will be enhanced, as resources permit, by increasing the university participation in space life sciences programs and increasing educational opportunities.

2.   Maintain approved ground-based projects (those in design, development, and operations, Phase C/D/E, respectively).

3.   Maintain definition of the appropriate flight facilities through Phase B.

4.   Maintain the necessary advanced planning studies through Phase A and Pre-Phase A.

FY 1992 will be a critical year for the space life sciences program. The requested budget would have provided a stable base for the life sciences and allowed for a progressive implementation of the research and development program, both ground and flight. It is imperative that the resulting FY 1992 budget be judiciously applied to the maintenance of ongoing core research activities at a sustainable level, to the extent possible, while supporting SSF planning and associated research, and enhancing, where possible, research to support MFPE.

A.   BUDGET

Using the guidelines and decisions described above, a set of options is being developed which will allow the implementation of previously described strategies. These guidelines provide for the application of a consistent decision process concerning programmatic priorities at the disciplinary level when faced with the realities of the overall Federal budget. The commitment of the Administration, the U.S. Congress, and NASA to the implementation of the overarching missions and the National Space Policy, while maintaining a viable research and development base, will have a profound impact on the execution and schedules of the space life sciences program. Another major factor in the implementation of the program will be the participation and contributions of other U.S. agencies, such as NIH, NSF, DOE and DOD, and of international partners in the space life sciences.

The budgetary guidelines developed by the space life sciences program are aimed primarily at protecting and preserving the core program from potential impacts and subsequent delays which might occur in other major flight projects and/or overarching initiatives. Thus the core program will continue to accumulate the knowledge base by maintaining and supporting a viable life sciences research and bioengineering community within and outside of NASA. This knowledge base is also essential to the success of NASA in maintaining the life sciences infrastructure when new initiatives and major missions are implemented.

While the preservation of the core program is of highest priority for the successful implementation of the space life sciences program, the core program cannot be preserved under the current budgetary environment. An approximately flat budget in Research and Analysis poses a major near-term challenge. For example, a decline in small grant programs has been necessary in 1992. Much work has been deferred. An option under consideration is to enhance the growth of the life sciences program in FY 1993 commensurate with the overall agency's Space Shuttle missions and SSF schedules, thereafter maintaining the core program at the necessary annual growth rate of 14 to 15 percent after inflation.

B.   ACCESS TO SPACE

Most of the missions currently being planned utilize the Space Shuttle and Spacelab missions, bioplatforms, and SSF. International missions utilizing both U.S. and foreign satellites are an important part of the space life sciences program. The Life Sciences Division also facilitates and integrates flight research sponsored by other U.S. Government agencies. The implementation of this strategy has resulted in the NASA life sciences programs developing flight experiments for unmanned Soviet COSMOS biosatellite missions, Space Shuttle middeck locker capabilities, and a variety of Spacelab missions, both dedicated to life sciences and in concert with other participating U.S. organizations and international partners. The Spacelab missions now enable the performance of experiments that require 7 to 10 days of flight and human intervention on-orbit. A limited range of analytical capability is provided on these missions, with the primary objective being to perform observations and collect samples and specimens during flight for subsequent Earth-based analyses. The extension of Spacelab flights to periods of time beyond 13 days will enable experiment replication on the same flight.

In the near term, the continuing Space Shuttle/Spacelab series of missions will be the most important categories of flight opportunities. It is essential that the extensive life sciences research conducted on Earth be accorded continuing and regular access to flight as a means of collecting unique data, validating scientific concepts and procedures, and verifying countermeasures, while serving as a testbed for future SSF research techniques and equipment. While the SLS-1 mission flown in June 1991 provided significant increases in the quantity and quality of research data in the Life Sciences archives, more questions were raised than definitively answered.

C.   TECHNOLOGY

A strong ongoing program of technology development must be pursued to enable space life sciences activities.

To support the planned research thrusts utilizing Space Shuttles, Spacelabs, SSF, and biosatellites, automated monitoring control and analysis capabilities are key. Also of prime importance are centrifuge facilities, animal habitats and supporting equipment, refrigerator-freezers, and sample processing and delivery techniques and equipment. Ground-based and observatory science needs include, for example, enhanced signal processing and detection systems for SETI, and telescience capabilities.

Technology requirements for current or planned human missions include improved EVA and life support systems, real-time environmental control, radiation monitoring, automated expert systems for medical care, microbial monitoring and decontamination. Mission from Planet Earth presents a significantly greater need for advanced technologies to enable human exploration missions. These include radiation shielding, regenerative life support systems, and automated and expert medical care and delivery systems. Artificial gravity systems may be required for exploration missions; these technologies could take the form of tethered or rigid systems, providing partial or continuous gravity as a countermeasure.

D. INSTITUTIONS

The Life Sciences Division will continue to develop an infrastructure to facilitate the future program growth required to support the major exploration initiatives of the 21st century. The basic structure is in place, and will be enhanced by upgrading NASA field center facilities and capabilities, enriching cooperative programs with other Federal agencies, strengthening ties to the university and industrial communities, and increasing international collaboration.

1. NASA Infrastructure

NASA Headquarters. The Life Sciences Division of the Office of Space Science and Applications is responsible for the overall planning and direction of NASA's space life sciences program. Within the Life Sciences Division, three program branches and an Aerospace Medicine Office are responsible for the scientific, technical and programmatic aspects of the Division's activities. Other staff elements include the Chief Scientist and Strategic Planning and Program Control. The Life Sciences Division is responsible for implementing the programs described in this document, and coordinating with other NASA Headquarters organizations involved in life sciences/life support activities — the Office of Aeronautics and Space Technology for physical-chemical life support systems and other life support hardware advanced development; the Office of Space Systems Development for life sciences/life support hardware development and mission operations and utilization planning; the Office of Space Flight for crew health operational issues; and the Office of Exploration in planning for the Space Exploration Initiative.

NASA Field Centers. NASA field centers are the source of expertise for program definition, research program implementation, unique ground and flight facilities, and flight mission planning and operation. A NASA Administrator's decision memorandum, dated December 30, 1991, based upon recommendations contained in the NASA "Roles and Missions" Report authorized by the NASA Deputy administrator, designated "Centers of Excellence (Science)" as follows:

ARC — Life Sciences (Gravitational biology, CELSS, exobiology)
JSC — Life Sciences (Humans — human physiology, operational and clinical medicine)

The following summarizes the major life sciences project management responsibilities assigned to NASA field centers:

- Ames Research Center— Space Biology Initiative (SBI—nonhuman research, Gas-Grain Simulator, CELSS Test Facility, Gravitational Biology Facility); Centrifuge Facility Project; Rhesus Facility; Research Animal Holding Facility; COSMOS flights; SETI and Life Support.

- Johnson Space Center—Operational medicine; SSF, CHeCS; EDOMP; BMAC; SBI (human research elements, Space Physiology Facility); Cosmic Dust Collection Facility; Spacelab Life Sciences missions; and human research on Mir.

- Marshall Space Flight Center—SSF Environmental Control and Life Support System. Management for Spacelab missions.

- Kennedy Space Center—CELSS Breadboard Project and the Space Life Sciences Training Program.

Advisory System. The life sciences advisory infrastructure is an extremely important tool for program assessment and long-range planning; it encompasses both internal and external committees. The Life Sciences Division will continue to receive advice from National Academy of Sciences committees focusing on life sciences disciplines. The NASA internal advisory infrastructure includes three committees— the Aerospace Medicine Advisory Committee (AMAC) in the area of clinical and preclinical medical studies necessary to enable human space flight, and the Space Science and Applications Advisory Committee (SSAAC) and its Life Sciences Advisory Subcommittee (LSAS), primarily in the area of nonmedical basic science studies.

2. Universities

The university community continues to offer a broad research and training base for the Life Sciences Division. This diverse community is also the source of the majority of the members of the discipline working groups and advisory elements. The initiation of the NASA Specialized Centers of Research and Training (NSCORT) Program and the establishment of the first centers in 1990 and 1991 are highly significant steps in the strengthening of university-based research in life sciences. As NASA's planning for exploration missions matures and resources become available, university research will assume even greater importance in select areas of research.

3. Other Federal Agencies

Joint activities with other Federal agencies are essential to the continued development of a national space life sciences program base. The Interagency Working Group (IAWG) in biomedical research will continue to enhance joint efforts between NASA and the National Institutes of Health (NIH). Complementary ground-based research programs are being developed in the cardiopulmonary, musculoskeletal, and neuroscience disciplines. Similar cooperative programs are planned with the Departments of Agriculture, Energy, and Defense, and the National Science Foundation. These joint activities will ensure that ground-based and flight facilities, with the appropriate management and scientific infrastructure, are available to support the needs and interests of other agencies as well as those of NASA in the utilization of space as a scientific tool. The Life Sciences Division will coordinate across the spectrum of organizations interested in space life sciences research. Specific interests will be integrated to ensure the development and execution of ground and flight research programs that meet the needs of all participants.

4. Ground-Based Facilities

A significant portion of NASA space life sciences research is conducted in ground-based laboratories. These investigations are conducted for the purposes of validating space-flight data; simulating conditions that might occur in space flight; developing analog models for weightlessness; and verifying new hypotheses prior to the execution of flight experiments. Over the last 15 years unique ground-based facilities were developed at NASA field centers and universities. The following represents a partial list of such facilities:

- Ames Research Center: Bedrest facility, human and animal-rated centrifuges, flight simulators, variable gravity research facility, biological research laboratories, crop growth research chambers, and research animal colonies.

- Johnson Space Center: Flight simulators, SSF health maintenance training facility, clinical research laboratory facilities, proportional radiation counter facility, flight baseline data collection facility, hypo- and hyperbaric human-rated facilities, water immersion training facility, human factors facility, life support test chambers, toxicology facility, and Spacelab Life Sciences and CHeCS training facilities.

- Kennedy Space Center: Preflight baseline data collection facility and CELSS breadboard facility.

- Universities: Slow rotating room at Brandeis University.

- National Laboratories: The Lawrence Berkeley Laboratory BEVALAC and the Booster Synchrotron at Brookhaven National Laboratory.

E. INTERNATIONAL COOPERATION

The Life Sciences Division is continuing to pursue a vigorous international program involving the major space agencies of the world, especially in the training of life scientists. The goal of international cooperation in the space life sciences is to increase the overall worldwide science return from space life sciences research. Coordinated activities are underway with Canada (CSA), the European Space Agency (ESA), the Federal Republic of Germany (DARA), France (CNES), Japan (NASDA), and Russia. In addition, discussions have been held and will continue with several other countries to determine the feasibility of establishing formal collaborative relationships.

An International Life Sciences Strategic Planning Working Group was established in 1990. The Life Sciences Division is participating with international partners in the planning and development of Spacelab International Microgravity Laboratory (IML) missions. SL-J is a Japanese Spacelab mission and the SL-D is a series of dedicated German Spacelab missions in which NASA Life Sciences participates. France, through CNES, is participating in the development of the Rhesus Facility, which will be flown on the Spacelab Life Sciences 3 mission. Five partners—France, Europe, Japan, Germany, and Canada—have been actively planning for science experiments to be flown on biosatellite missions. Mechanisms for joint experiments and sharing of data from U.S. and Russian manned and unmanned missions have been established. In planning for the utilization of SSF for life sciences research, international partners intend to provide complementary inflight facilities, and joint studies will be conducted.

APPENDIX I: SPACE LIFE SCIENCES PROGRAMS — INDIVIDUAL STRATEGIES

- Life Support Programs

  Space Physiology and Countermeasures
  Radiation Health
  Environmental Health
  Space Human Factors
  Controlled Ecological Life Support Systems

- Research Programs

  Space Biology
  Biospheric Research
  Exobiology

- Operational Medicine Program

PROGRAM GOALS

The Space Physiology and Countermeasures Program has two goals:

- Understand the underlying mechanisms of the physiological changes that occur during space flight, and

- Develop and validate countermeasures and technologies to optimize crew safety, well-being, and performance inflight and on return to Earth.

VISION

By the end of the century, as a result of our programs of flight and ground-based research, we will have achieved significant progress in understanding the mechanisms underlying physiological adaptation to space flight and in developing countermeasures to the undesirable effects of spaceflight on crewmembers.

We will emphasize application of space physiology research and technology to disease and other health problems on Earth.

LEVEL I SCIENCE REQUIREMENTS

Develop systems and procedures to maintain crew health and performance sufficient to accomplish mission objectives. This will include the following:

- Cardiopulmonary Physiology - Understand mechanisms of changes in cardiovascular function with space flight and develop countermeasures using real and simulated weightlessness in humans and animals.

- Musculoskeletal Physiology - Understand mechanisms, determine health consequences, and develop countermeasures to muscle atrophy and bone demineralization using real and simulated weightlessness. Develop exercise equipment and procedures for maintaining bone integrity and muscle strength.

- Neuroscience - Understand vestibular and sensory adaptation, central nervous system processing, and neuro-motor control in weightlessness and develop countermeasures where appropriate, including effective countermeasures to space motion sickness.

- Regulatory Physiology - Understand integrative mechanisms regulating responses to space flight, including those related to circadian rhythms, fluid and electrolyte balance, endocrinology, pharmacodynamics, metabolism and nutrition, immunology, hematology, and temperature regulation, and develop requirements and countermeasures where necessary, including nutritional requirements for long-duration missions.

THE STRATEGY

Research in Space Physiology and Countermeasures to accomplish these objectives over the next decade will build on the existing science knowledge base and will consist of a combination of integrated ground-based and flight research.

- Animal models will be important in investigating basic mechanisms of physiological adaptation.

- Human subjects, including crewmember volunteers for Spacelab and other Space Shuttle missions and volunteer subjects for ground-based research using simulations of space flight, will be important in investigating applied research issues.

Collaborative efforts with our international partners will include continued cooperation with the Soviets to share equipment and data in their Mir Program and to fly experiments on the Cosmos biosatellite; collaboration with the French and German Space Agencies on flight and ground research; cooperation with the Japanese and European Space Agencies on Spacelabs and SSF; and cooperation with the Canadians on Spacelabs.

NEW INITIATIVES

- Neurolab (Spacelab)
  In connection with the U.S. Interagency Program "Maximizing Human Potential - Decade of the Brain 1990-2000," NASA will have a Spacelab dedicated to brain and behavior, in which state-of-the-art neurosciences research will be conducted.

- NSCORT Program in Integrative Physiology.

PROGRAM GOAL

The overall goal of the Radiation Health Program is to establish the scientific basis for the radiation protection of humans engaged in the exploration of space, with particular emphasis on lunar and Mars missions.

VISION

By the end of this century the Radiation Health Program will have enabled us to predict the probabilities, in excess of natural incidence, of deleterious health effects, including carcinogenic effects due to radiation exposure.

LEVEL I SCIENCE REQUIREMENTS

SHUTTLE:
- Monitor radiation exposures to crews.

SPACE STATION FREEDOM:
- Monitor radiation exposures to crews to assure that guidelines for radiation exposure limits are being met.

SPACE EXPLORATION INITIATIVE:
- Determine health risks from GCR and develop appropriate exposure limits.

- Determine health risks for exposures to solar energetic particles and develop warning and shielding requirements for solar particle events.

THE STRATEGY

The strategy to meet the science requirements consists of a ground-based research program to develop the scientific basis for radiation protection in space, complemented by a space-based program to validate the scientific results in the space radiation environments and inside the spacecraft and other structures. Some of the studies required to meet these requirements are properly performed by other components of NASA, other Federal agencies, Federally supported research laboratories, universities and foreign institutions.

The ground-based research program will be executed at one or more accelerator facilities, which provide beams of protons and heavier charged particles to simulate the space radiation environment. Reference beam species and beam energies will be used to integrate experiments and enable intercomparison of their results. The biological research program focuses on molecular and biological studies to elucidate the mechanisms of cellular responses to radiation, and on tissue and organ studies to elucidate the cellular response, with particular reference to life-shortening effects, mainly cancer.

The space-based program is being redefined in order to develop adequate tools for space verification of ground-based predictions on a cost-effective platform, using multiparameter studies to test sensitive aspects of the predictive models that require improvement. Some of the predictions that will be tested with statistically significant results include the relative abundance of different particles and their flux behind different thicknesses of test materials, transformation of human cells in culture, differential genetic responses to radiation of different ionization density in small organisms, and effects of radiation and weightlessness on the development of microorganisms and plant seeds. The outcome of the comparison of ground-based prediction with space-based experiment is a measurement of the uncertainty associated with the prediction of radiation effects in space. This is the most important product of the research program.

NEW INITIATIVES

- Radiation Biology Initiative. This will develop and operate facilities on the ground (either the Lawrence Berkeley Laboratory or the Brookhaven National Laboratory) and in space (LifeSat) for heavy ion radiobiology.

- NSCORT Program in Radiation Health.

LIFE SUPPORT
ENVIRONMENTAL HEALTH

PROGRAM GOALS

The Environmental Health Program, which comprises microbiology, toxicology, and barophysiology, has three goals:

- Utilize ground-based studies to understand the effects of the spacecraft and EVA environments on humans.

- Specify, measure, and control these environments.

- Develop countermeasures where necessary to optimize crewmembers' health, safety, and productivity.

VISION

The Environmental Health Program by the end of the century will produce an understanding of the environmental hazards within and outside of spacecraft and will have developed effective monitoring and countermeasures to assure the safety of crewmembers on long-duration missions. Applications for humans on Earth will be emphasized.

LEVEL I SCIENCE REQUIREMENTS

1. Define microbiological and toxicological standards and develop advanced environmental monitoring technology.

2. Understand the biochemical and biophysical effects of variations in component parts of the man-made atmosphere in space environments to allow selection and maintenance of safe and efficient gaseous environments in different situations and in different eras of the space program.

3. Determine whether a prolonged stay in microgravity would induce sufficient physiological and biochemical changes to render crewmembers more susceptible to toxic chemicals.

4. Establish acceptable and appropriate ranges of gas composition, pressure, temperature, and humidity for all current and future missions.

5. Determine the interactive effects of all potential atmospheric components and factors on physical and psychological well-being and crew performance.

6. Determine the effect of microgravity on the microbial cell, its growth, metabolism, pathogenic potential, virulence factors, and sensitivity to antibiotics and disinfectants.

7. Establish criteria for the number of microbial organisms in a defined volume of air and water that could be considered realistic and safe.

8. Define safe protocols for training and define decompression models to predict assessment of the incidence of decompression sickness.

THE STRATEGY

Research in Environmental Health to accomplish these goals will build on existing *science and operational* knowledge bases and will involve an integrated ground and flight research effort. The Neutral Buoyancy Laboratory (NBL) and chamber tests simulating EVA decompression will be used to define safe protocols for training and decompression, expand the data base, and develop models.

Routine monitoring of Space Shuttle flights for selected toxicological contaminants and research on microbial contamination and growth will be conducted.

The Environmental Health NSCORT at the University of Rochester will be synergistic with other aspects of the Environmental Health Program and will identify hazardous situations, develop detection systems, and use data to develop computer models to estimate growth and flow of contaminants.

NEW INITIATIVES

Ecolab Spacelab. A Spacelab dedicated to all aspects of the spacecraft environment, including environmental health issues, will be investigated.

LIFE SUPPORT
SPACE HUMAN FACTORS

PROGRAM GOALS

The goals of the Space Human Factors Program are to:

• Understand the psychological, behavioral, and performance responses to space flight.

• Develop design requirements, protocols, and countermeasures to enable safe, productive, and enhanced crew performance.

VISION

By the end of the century, through a combination of flight and ground-based research, the Space Human Factors Program will have made significant progress in understanding crew needs and developing systems in the areas of crew support, small-group interactions, mission work analysis, workload assessment and performance, selection and training, habitability, human-machine systems and automation.

LEVEL I SCIENCE REQUIREMENTS

It is important to develop the knowledge base required to understand the basic mechanisms underlying behavioral adaptation to space flight and the capabilities and limitations of the crewmember in the unique environments that will be encountered during future space missions. This will include:

• Understand behavioral processes and performance capabilities in space.

• Develop habitability requirements for extended-duration missions.

• Develop guidelines for human-machine interactions for extended-duration missions.

• Develop crew support systems for extended-duration missions.

• Develop selection and training protocols for extended-duration missions.

• Develop principles and requirements for crew organization, leadership, and composition for extended-duration missions.

• Develop human performance requirements for operations and procedures for space missions.

• Develop a glove to allow dexterity in EVA.

THE STRATEGY

To fulfill these science requirements, the Space Human Factors Program will build upon the existing knowledge base, which is an extensive and comprehensive core of ground-based and aviation-based research and technology. The ground-based program will use analogs, mock-ups, laboratory studies, modeling and simulations to study psychosocial factors important to crew health, well-being, and performance.

Human performance experiments are planned for the International Microgravity Laboratory 2 (IML-2) scheduled for flight in 1994.

In addition, interagency cooperation will be developed within the program. The Space Human Factors Program will work with the U.S. Air Force in the areas of crew workload and mission analysis; with the National Oceanographic and Atmospheric Administration (NOAA) to use the undersea habitat as an analog for studying effects of confinement on psychological health and group dynamics; and with the National Science Foundation (NSF) to develop the Arctic and Antarctic as analogs for space exploration.

International collaborative projects include the Rhesus Research Facility currently being developed by NASA and the French Space Agency Centre Nationale d'Etudes Spatiales (CNES) to study behavioral processes in microgravity. Similar studies will be conducted in conjunction with the Soviets during joint activities on Cosmos biosatellites, and crew selection studies are planned on the Mir space station. The Space Human Factors Program plans collaboration with the German Space Agency (DARA), and is negotiating with the European Space Agency (ESA) on potential joint participation in isolation studies.

NEW INITIATIVES

Initiatives being proposed within the Space Human Factors Program during the next 5 years include:

- The Human Factors Analog Initiative. Initiative to utilize isolated, confined and potentially hazardous analogs to study selected space physiology issues in the polar regions (Antarctic and Arctic) and in undersea habitats.

- NSCORT Program in Human Factors.

PROGRAM GOAL

The goal of the CELSS Program is to develop regenerative life support systems by combining biological and physical/chemical processes capable of producing and recycling the food, air, and water needed to support long-term human missions in space in a safe and reliable manner.

VISION

We will design, develop, and test a ground-based human-rated CELSS and simultaneously initiate a space-based experiment program to identify and resolve issues related to the behavior and operation of CELSS biological and non-biological subsystems by flight experimentation in micro- and fractional-gravity. Ultimately, a human-rated, space-based CELSS will provide enabling life-support capabilities for long-duration missions.

LEVEL I SCIENCE REQUIREMENTS

To guide the research and technology development needed, three supporting research goals have been established:

1.   Understand how life can be maintained in stable, autonomous systems.

2.   Acquire the knowledge needed to make life support systems for long-duration   missions independent of resupply.

3.   Develop the technology base needed to build autonomous life support systems.

THE STRATEGY

The approach to the development of the CELSS will be to emulate the fundamental mechanisms by which life is sustained on Earth. Three essential terrestrial ecological processes are photosynthesis, respiration, and microbial mineralization. However, because of the constraints imposed by space missions, i.e. limited volume, mass, energy, and human resources, CELSS will need to substitute physical/chemical and computational processes or devices for the natural biological and geological feedback mechanisms that operate on Earth. The overall goal will be to develop systems that will provide life support consumables in an effective, efficient, safe, and reliable manner within anticipated mission constraints.

To carry this out, the following strategy will be followed:

1.   Subsystem Development

Carry out small-scale laboratory-based research and technology development on individual biological and non-biological systems and processes of the CELSS to select the best candidate subsystems for inclusion in CELSS systems.

2.   System Studies

Develop increasingly sophisticated CELSS prototype systems that display sequentially greater size, degree of integration, and level of complexity. These facilities will include small-scale integrated subsystems, and progress to breadboard systems, ground-based human-rated testbeds, and eventually evolve in engineering prototypes of space CELSS for a specific mission application. Conceptual and mathematical modelling studies will be included in these system studies.

3.   Flight Studies

Perform flight experiments that test the effect of the total space environment on CELSS design features and processes (including physical/chemical and biological). This will include gravity, radiation sensitivity, and other parameters of the space environment.

This strategy will be implemented by maintaining close collaboration and cooperation with relevant offices both within and outside of NASA. The CELSS program will benefit from the expertise offered by such NASA programs as the Space Biology Program of the Life Sciences Division, the Advanced Life Support Program of the Office of Exploration and Space Technology, and the ECLSS Program of the Space Station Freedom Program Office. Outside NASA, the U.S. Department of Agriculture, the U.S. Department of Energy, and the National Science Foundation will provide needed information of basic and applied plant and crop research. The industrial and commercial sectors may also be involved in research concerning various aspects of the CELSS program.

4.   NSCORT

The major focus of the NASA Specialized Center of Research and Training (NSCORT) at Purdue University is the parallel development of food production, food processing, and waste management for a space-operated controlled ecological life support system (CELSS). An interdisciplinary group with expertise that ranges from systems engineering to biotechnology will (1) apply recombinant DNA techniques to appropriately modify photosynthetic microorganisms and crop plants for the food, atmospheric, and energy requirements of a CELSS; (2) specify human nutritional requirements under long-duration missions and for colonization under a hypogravity environment; and (3) model the overall life support system in order to integrate all important subsystems of a functioning CELSS. The overall research effort will be complemented by an extensive training component that will include postdocs, graduate students, and undergraduates.

NEW INITIATIVES

Human-rated facilities for CELSS development.

RESEARCH PROGRAMS
SPACE BIOLOGY

Program Goal

To advance biological knowledge and understand gravity's role in biological processes by using microgravity as a research tool.

Vision

A scientific understanding of the effects of gravity on plants and animals at the cell and molecular level will be achieved, and new basic knowledge of biological processes will contribute to accomplishing NASA's goals and improving the quality of life on Earth.

Level I Science Requirements

Research will be conducted on simple to complex plants and animals and cells to achieve the following scientific objectives in diverse representative species:

1.   Gravity Sensing/Perceptions Identify gravity-sensing organs and mechanisms; and understand how g information is transduced, processed, transmitted and integrated into a response.

2.   Development and Growth Determine if multiple generations can develop normally in microgravity; identify and understand effects of gravity and microgravity on gravity-sensitive developmental stages, systems and mechanisms; and determine the role of gravity in development and evolution.

3.   Integrative Biology Identify the effects of gravity on the comparison, regulation and function of biological support structures; identify the role of gravity in regulating metabolism, fluid dynamics and biorhythms; and identify effects of the interaction of quantity with other environmental factors.

4.   Cell Biology Determine how and where gravity effects cells; distinguish direct from indirect gravitational effects on cells; identify the role of gravity in maintaining normal cellular function; and assess the permanence of microgravity effects on cells.

The Strategy

•   Develop an augmented academic community in gravitational biological research

•   Train the next generation of space biologists who will use Space Station Freedom

•   Expand support of ground-based research utilizing modern techniques and technology to target analyses at the most fundamental levels.

•   Continue to develop a vigorous flight experiments program using the full range of available flight opportunities to permit identification, confirmation, and understanding of effects due to microgravity, and to generate a scientifically valid data base.

•   Develop in-house science capability, including instruments, facilities and personnel support

•   Expand cooperative efforts with action of NASA programs, international collaborators, and other government agencies to maximize utilization of resources and generation of scientific data

New Initiatives

- A Gravitational Biology Enhancement has been put forth to garner the funds required to implement the Space Biology Program Strategy

- In preparation for Space Station Freedom's utilization both the Centrifuge Facility and Space Biology Initiative Gravitational Biology Facility have been approved and funded

- A new NSCORT in plant biology has been proposed

PROGRAM GOAL

To utilize ground and space-based studies to understand how biological processes and planetary properties affect one another, and how human activities affect this interaction

VISION

We will obtain a predictive understanding of the mechanisms by which the terrestrial biosphere interacts with the planet, how anthropogenic activities can perturb these interactions, and will define strategies by which these perturbations can be moderated or mitigated.

LEVEL I SCIENCE REQUIREMENTS

1. Establish a basis for assessing the major pathways and rates of exchange for carbon, nitrogen, sulfur, and phosphorus moving into and out of terrestrial and aquatic ecosystems.

2. Establish a basis for extrapolating local rates of an biological activities to biospheric rates and effects, with particular attention on the role of gases and their oxidation products.

3. Develop mathematical models that represents the dynamics of the global cycles of carbon, nitrogen, sulfur, and phosphorus, including their interactions and the key processes that control their dynamics.

4. Study mechanism of population and community ecology as underlying biogeochemical processes.

5. Apply these insights and technologies to issues of global human health and welfare.

THE STRATEGY

In order to more fully accomplish the above programmatic goals and objectives, discussions are underway between the Life Sciences Division and the Earth Science and Applications Division concerning the transfer of program elements in FY 1993. This management change is directed toward more closely integrating the Biospheric Research Program with the major NASA programs in Earth observations and global ecology. The Life Science Division will continue to manage global monitoring and disease monitoring activities utilizing remote sensing technology.

RESEARCH PROGRAMS
EXOBIOLOGY

PROGRAM GOAL

The goal of NASA's Exobiology program is to understand the origin, evolution, and distribution of life in the universe. The attainment of this goal is being sought by concentrating on specific research objectives that trace the pathways taken by the biogenic elements, leading from the origin of the universe through the major epochs in the evolution of living systems and their precursors. These epochs are 1) The cosmic evolution of the biogenic compounds, 2) prebiotic evolution, 3) the early evolution of life, and 4) the evolution of advanced life.

VISION

The Exobiology program conducts research as part of the Life Sciences Division program, and interacts with programs elsewhere within NASA in an integrated effort to accomplish the goal of understanding life in the universe.   Exobiology investigations formed an integral part of the science objectives of the Solar System Exploration Division, and also formed a major portion of the National Academy of Sciences recommendations for research in Astronomy and Astrophysics. Accordingly, advanced planning within the Exobiology program encompasses joint activities with both the Solar System Exploration Division and the Astrophysics Division within OSSA. Future plans include research activities conducted on the ground, aboard robotic planetary spacecraft and Earth-orbiting telescopes, on the Space Shuttle and on Space Station Freedom, and eventually research conducted by humans on or from the Moon and Mars.

LEVEL I SCIENCE REQUIREMENTS

1.   Determine the history of the biogenic elements (C, H, N, O, P, S) from their birth in stars to their incorporation into planetary bodies.

2.   Understand the pathways and processes leading from the origin of a planet to the origin of life. Research strategies in this area investigate the planetary and molecular processes that set the physical and chemical conditions within which chemical evolution occurred and living systems arose.

3.   Determine the nature of the most primitive organisms, the environment in which they evolved, and the way in which they influenced that environment by investigating two natural repositories of evolutionary history available on Earth and perhaps elsewhere: the molecular record in living organisms and the geological record in rocks.

4.   Determine the extrinsic factors influencing the development of advanced life and its potential distribution, including an evaluation of the influence of extraterrestrial influences and planetary processes on the appearance and evolution of multicellular life.

THE STRATEGY

•   Maintain a broad program of investigations of the origin, evolution, and distribution of life on Earth over the course of Earth's history, in order to understand the context of life in the universe.

•   Prepare exobiologists to make the maximum use of NASA and foreign flight opportunities to extend the range and quantity of space measurements appropriate to exobiology.

•   Develop a suite of exobiology capabilities and instruments for inclusion on future planetary missions to the inner and outer planets, and to the small bodies of the solar system.

•   Use Space Station Freedom as a platform to conduct fundamental experiments to characterize the role of the biogenic elements and compounds in the origin and evolution of the solar system and its constituents.

- Prepare for the scientific exploration of the Moon and Mars by a combined program of analog studies, instrument development, and robotic and human exploration.

- Conduct observational exobiology using all available platforms to study the formation of life in the universe, and to search for signs of extraterrestrial life.

NEW INITIATIVES

The Program is preparing for the observational phase of the SETI Microwave Observing Project (1992), the Mars Observer mission (1992), the USSR Mars '94 mission (1994), the arrival of the Galileo mission at Jupiter and the launch of the CRAF and Cassini missions to a comet and the Saturn system, and the installation of the Gas-Grain Simulator on Space Station Freedom. Steps are also being taken to prepare for future Space Exploration Initiative missions to Mars and later the Moon. One early step will be the initiation of joint research with the National Science Foundation (NSF) in Antarctica in the 1992-93 season. Follow-on activities in SETI and solar system exploration are planned for initiation early in the first decade of the 21st century.

OPERATIONAL MEDICINE

PROGRAM GOALS

The goals of the Operational Medicine Program are as follows:

- Provide medical support to all manned missions,

- Develop an estimate of medical risks associated with advanced missions with humans,

- Establish biomedical countermeasures and life support research priorities.

VISION

Two primary missions will shape operational medicine activities in the coming decade. The Shuttle Program, under the sponsorship of the Extended Duration Orbiter Program, will fly progressively longer missions during the first half of the 1990s. The EDO Medical Program (EDOMP) is tasked with supporting missions up to 16 days, while assuring the safety of the environment and the capability of crewmembers to function effectively during re-entry, landing, and land contingencies. The key challenge facing the EDOMP is to develop (and validate) effective countermeasures, and to routinely support missions of 13-16 days. To accomplish this task, it will be mandatory to acquire the medical/physiological knowledge to understand the mechanisms of deconditioning on missions of this duration. The ultimate objective is to develop the ability to predict a given crewmembers response to re-entry stresses by doing on-orbit evaluations.

Space Station Freedom (SSF) embodies the U.S. commitment to a permanent presence in low-Earth orbit. The principal aeromedical challenge embodied in the SSF program is the progressive extension of crew stay times in microgravity. Experience to date in both the U.S. and Soviet space programs has resulted in the identification and initial characterization of the human response to long-duration space flight. While many of these responses are adaptive in nature, certain adaptations are maladaptive and may result in medical illness (either inflight or postflight), or impair a crewmember's ability to perform critical mission tasks. These processes must be understood well enough to develop predictive models and methods for on-orbit assessment of deconditioning trends. The programs required to accomplish these objectives will entail a high degree of synergy with other ground-and flight-based research programs, both of an applied and a basic nature. The changing nature of the space program also results in two additional challenges - first, the need to better characterize the medical risks of routine, repetitive EVA, and second, a need to provide quality medical care to crews in an Earth-remote location. A well-designed medical care capability has a role in both preventing morbidity and mortality, and in keeping minor medical problems from progressing to the extent that they result in mission replanning or a medical abort.

In the latter half of this decade, the Operational Medicine Program will dedicate increasing resources to advanced medical concepts for exploration missions. Specific elements will include: 1) requirements definition for third generation medical care systems (including computer-aided diagnostic and telemedicine support), 2) development of crew selection and medical standards, and 3) development of procedures and protocols for health care delivery during long-duration missions. This latter task will require an understanding of the medical consequences of illness/injury states in the space-adapted individual.

LEVEL 1 MEDICAL REQUIREMENTS

1. Understand the medical consequences of spaceflight deconditioning and their operational implications.

2. Characterize the operational medical risks of the proposed missions, including the physiological risks of repetitive EVAs, and the potential medical risks associated with the Shuttle EDO and SSF environment.

3. Develop an understanding of the efficacy of currently employed medical countermeasures and establish biomedical research priorities.

4. Develop the requisite understanding of medical physiology in the space-adapted and/or ill injured crewmember in order to allow medical care in space, which is based on that underlying knowledge.

5. Development of advanced medical care systems and telemedicine concepts to deliver quality medical care in space, within the context of programmatic constraints.

THE STRATEGY

The strategy for meeting these objectives is as follows:

- Develop an understanding of the mechanisms of spaceflight deconditioning by implementing a preventive medicine program, which will limit the progression of those effects. This will be accomplished by direct inflight crew monitoring (such as in the EDOMP and BMAC programs), or by an active interface with the applied life sciences research programs (such as Spacelabs SLS-1, 2, & 3).

- Implementation of the preventive medicine program by means of ground-based medical care and an inflight element which includes routine clinical examinations and prescribed countermeasures to prevent illness/debility due to spaceflight deconditioning. This function is represented programmatically within the Crew Health Care System (CHeCS.)

- Monitoring of spacecraft air, water, and cabin surfaces are an important element of the preventive medicine program. This activity (for SSF) will be implemented by a subsystem (the Environmental Health System, or EHS) of CHeCS.

PLANNED INITIATIVES

Develop a program of clinical investigations into the effects of space flight adaptive mechanisms on the health status of sick or injured crewmembers, and methods for delivering medical care.

APPENDIX II: REFERENCE DOCUMENTS

1. 1979 — *Life Beyond the Earth's Environment: The Biology of Living Organisms in Space* (The Bricker Report), Space Science Board, National Research Council

2. 1981 — *Origin and Evolution of Life-Implications for the Planets: A Scientific Strategy for the 1980's* (The Margulis Report), Committee on Planetary Biology and Chemical Evolution, National Research Council

3. 1988 — *Earth System Science: A Closer View* (The Bretherton Report), Earth System Sciences Committee, NASA Advisory Council

4. 1987 — *Space Technology to Meet Future Needs* (The Shea Report), Committee on Advanced Space Technology, Aeronautics and Space Engineering Board, Commission on Engineering and Technical Systems, National Research Council

5. 1987 — *A Strategy for Space Biology and Medical Science for the 1980's and 1990's* (The Goldberg Report), Committee on Space Biology and Medicine, Space Science Board, Commission on Physical Sciences, Mathematics, and Resources, National Research Council

6. 1988 — *Space Science in the Twenty-First Century: Imperatives for the Decades 1995 to 2015, Life Sciences* (The Swisher/Usher Report), Task Group on Life Sciences, Space Science Board, Commission on Physical Sciences, Mathematics, and Resources, National Research Council

7. 1988 — *Exploring the Living Universe: A Strategy for Space Life Sciences* (The Robbins Report), Life Sciences Strategic Planning Study Committee, NASA Advisory Council

8. 1986 — *Remote Sensing of the Biosphere* (The Botkin Report), Space Sciences Board, National Research Council

9. 1986 — *Global Change in the Geosphere-Biosphere: Initial Priorities for an IGBP*, National Research Council, National Academy of Sciences

10. 1990 — *The Search for Life's Origins: Progress and Future Directions in Planetary Biology and Chemical Evolution*, Committee on Planetary Biology and Chemical Evolution, Commission on Physical Sciences, Mathematics, and Applications, National Research Council

11. 1990 — *Report of the Advisory Committee on the Future of the U.S. Space Program*

12. 1991 — *America at the Threshold, Report of the Synthesis Group on America's Space Exploration Initiative*

13. 1991 — *Assessment of Programs in Space Biology and Medicine*, Committee on Space Biology and Medicine, Space Studies Board, Commission on Physical Sciences, Mathematics, and Applications, National Research Council

14. 1992 — *Vision 21 — The NASA Strategic Plan*, National Aeronautics and Space Administration